Essential Oils and Aromatic Plants

Difficile est scribere,
praesertim edenda

Essential Oils and Aromatic Plants

Proceedings of the 15th International Symposium on Essential Oils, held in Noordwijkerhout, The Netherlands, July 19–21, 1984

edited by

A. Baerheim Svendsen and J.J.C. Scheffer
Division of Pharmacognosy
Center for Bio-Pharmaceutical Sciences
Leiden State University, The Netherlands

1985 **MARTINUS NIJHOFF/DR W. JUNK PUBLISHERS**
a member of the KLUWER ACADEMIC PUBLISHERS GROUP
DORDRECHT / BOSTON / LANCASTER

Distributors

for the United States and Canada: Kluwer Academic Publishers, 190 Old Derby Street, Hingham, MA 02043, USA
for the UK and Ireland: Kluwer Academic Publishers, MTP Press Limited, Falcon House, Queen Square, Lancaster LA1 1RN, UK
for all other countries: Kluwer Academic Publishers Group, Distribution Center, P.O. Box 322, 3300 AH Dordrecht, The Netherlands

Library of Congress Cataloging in Publication Data

Library of Congress Cataloging in Publication Data

International Symposium on Essential Oils (15th : 1984 : Noordwijkerhout, Netherlands)
Essential oils and aromatic plants.

1. Essences and essential oils--Congresses.
2. Aromatic plants--Composition--Congresses.
I. Baerheim-Svendsen, A. II. Scheffer, J. J. C.
III. Title.
TP958.I57 1984 661'.806 85-11601

ISBN-13: 978-94-010-8772-8 e-ISBN-13: 978-94-009-5137-2
DOI: 10.1007/978-94-009-5137-2

CONTENTS

INTRODUCTION

The 15th International Symposium on Essential Oils was organized in Noordwijkerhout in the neighbourhood of Leiden, The Netherlands, also to celebrate that the very first beginning of these symposia took place in Leiden in the fall 1969. Then the pharmacognosists Dr. F.W. Hefendehl (Freiburg, FRG), Dr. K.-H. Kubeczka (Karlsruhe, FRG), Dr. J. Karlsen and Prof.Dr. A. Baerheim Svendsen (both Leiden, The Netherlands) came together to find out if it could be possible to organize quite informal meetings annually in order to discuss common problems concerning essential oil research. The meetings started fairly modest and the group was small, but it increased little by little, and later on more pharmacognosists and other people – from universities and from the industry – interested in essential oil research joined the group.

The 15th symposium was attended by about 80 participants from the following countries: Belgium, Egypt, England, Federal Republic of Germany, Finland, France, Hungary, Israel, Italy, The Netherlands, Portugal, Rwanda, Switzerland, Turkey and Yugoslavia.

So far, new methods and techniques in essential oil research have always been the main topics of the symposia; in addition reports on current research on essential oils and essential oil-bearing plants have been delivered.

The 15th symposium had as main topic 'headspace analysis of essential oils and aromatic plants', and the method as well as some applications of it were dealt with in a series of main and short lectures. Also other methods and techniques were discussed in the lectures. In a number of short lectures and in posters reports were given on current research on essential oils and their constituents as well as on essential oil-bearing plants, their propagation, cultivation, etc. A review was given on the biological effects and side-effects of essential oils and their constituents. 'The medicinal plants in the mirror of Dutch painting' was the title of the opening lecture.

We like to thank the authors for their contributions, and the following firms and organizations for their financial support:

Dragoco Gerberding & Co. GmbH, Holzminden, FRG
Johann Georg Fink GmbH & Co., Herrenberg, FRG
Givaudan Forschungsgesellschaft AG, Dübendorf-Zürich, Switzerland
Haarmann & Reimer GmbH, Holzminden, FRG
Hewlett-Packard Nederland B.V., Amstelveen, The Netherlands
Interpharm B.V., Farmaceutische Groothandel, 's-Hertogenbosch, The Netherlands
Naarden International N.V., Naarden, The Netherlands
NEA — Dutch Association of Flavours and Fragrances Manufacturers —, The Hague, The Netherlands
NEFARMA — Dutch Association of the Pharmaceutical Industry —, Utrecht, The Netherlands
OPG — Coöperatieve Apothekers Vereeniging 'De Onderlinge Pharmaceutische Groothandel' U.A. —, Utrecht, The Netherlands
Packard Instrument B.V., Delft, The Netherlands

A. Baerheim Svendsen J.J.C. Scheffer
Division of Pharmacognosy, Center for Bio-Pharmaceutical Sciences, Leiden University, The Netherlands

LIST OF CONTRIBUTORS

Abdou RF	Department of Genetics, Faculty of Agriculture, Assiut University, Assiut, Egypt
Aynehchi Y	Department of Pharmacognosy, School of Pharmacy, University of Tehran, Iran
Baerheim Svendsen A	see Janssen
Bishay DW	Department of Pharmacognosy, Faculty of Pharmacy, Assiut University, Assiut, Egypt
Blažević N	Chromos-Aroma, Zagreb, Yugoslavia
Brandauer H	see Ziegler
Brunke E-J	Research Division, Dragoco Gerberding & Co. GmbH, D-3450 Holzminden 1, FRG
Bruno M	Institute of Organic Chemistry, University of Palermo, Via Archirafi 20, I-90123 Palermo, Italy
Carmo MM	LNETI-DTIQ, Division of Natural Products, Estrada das Palmeiras, P-2745 Queluz de Baixo, Portugal
Coolsaet BA	see De Pooter
Czygan F-C	Institute of Botany and Pharmaceutical Biology, University of Würzburg, Mittlerer Dallenberg-weg 64, D-8700 Würzburg, FRG
Dirinck PJ	see De Pooter
Eley C	Department of Biology, University of Colorado, Colorado Springs, Colorado 80907, USA
Eley J	see Eley C
El-Keltawi NE	Department of Horticulture, Faculty of Agri-culture, Assiut University, Assiut, Egypt
Frazão S	see Carmo
Günther W	WGA, Senta-Weg 16, D-4000 Düsseldorf 30, FRG
Hammerschmidt F-J	see Brunke
Hanssen H-P	Institute of Pharmacognosy, University of Hamburg, Bundesstrasse 43, D-2000 Hamburg 13, FRG
Heide R ter	Research Department, Naarden International NV, P.O. Box 2, NL-1400 CA Naarden-Bussum, The Netherlands
Hiltunen R	Division of Pharmacognosy, School of Pharmacy, University of Helsinki, Fabianinkatu 35, SF-00170 Helsinki, Finland

Huber L	see Obbens
Huopalahti R	Department of Biochemistry, Laboratory of Food Chemistry, University of Turku, SF-20500 Turku 50, Finland
Husain SZ	Department of Botany, Plant Science Laboratories, University of Reading, Whiteknights, Reading RG6 2AS, England
Ikan R	see Weinstein
Janssen AM	Division of Pharmacognosy, Center for Bio-Pharmaceutical Sciences, Leiden University, P.O. Box 9502, NL-2300 RA Leiden, The Netherlands
Kesälahti E	see Huopalahti
Kolb B	Bodenseewerk Perkin-Elmer & Co. GmbH, P.O. Box 1120, D-7770 Uberlingen, FRG
Knobloch K	Institute of Botany and Pharmaceutical Biology, University of Erlangen-Nürnberg, Schlossgarten 4, D-8520 Erlangen, FRG
Kubeczka K-H	Institute of Pharmaceutical Biology, University of Würzburg, Mittlerer Dallenbergweg 64, D-8700 Würzburg, FRG
Kuftinec J	Laboratory of Chemical Pharmacology, Podravka Institute, Zagreb, Yugoslavia
Kuštrak D	Institute of Pharmacognosy, Faculty of Pharmacy and Biochemistry, University of Zagreb, Maruličev trg 20/II, P.O.Box 156, Yu-41000 Zagreb, Yugoslavia
Laakso I	see Hiltunen
Lamparsky D	Givaudan Research Company Ltd., Uberlandstr. 138, CH-8600 Dübendorf-Zürich, Switzerland
Lemberkovics É	Institute of Pharmacognosy, Semmelweis Medical University, Ullöi út 26/III, H-1085 Budapest VIII, Hungary
Mellerio G	Mass Spectrometry Center, University of Pavia, Via Taramelli 10, I-27100 Pavia, Italy
Mészáros S	see Verzár-Petri
Michaelis K	Institute of Organic Chemistry, University of Erlangen-Nürnberg, Henkestr. 42, D-8520 Erlangen, FRG
Obbens H	Hewlett-Packard GmbH, P.O.Box 1280, D-7517 Waldbronn 2, FRG
Paulini H	see Knobloch
Piozzi F	see Bruno
Pitarevič I	Dalmacijabilje, Dubrovnik, Yugoslavia

Pooter HL De	Laboratory of Organic Chemistry, Faculty of Agricultural Sciences, State University of Ghent, Coupure links 653, B-9000, Ghent, Belgium
Putievsky E	see Ravid
Ravid U	Division of Medicinal and Spice Crops, Agricultural Research Organization, Regional Experiment Station, Newe Ya'ar, P.O. Haifa 31-999, Israel
Şarer E	Faculty of Pharmacy, Ankara University, Tandogan-Ankara, Turkey
Schamp NM	see De Pooter
Scheffer JJC	see Janssen
Schilcher H	Institute of Pharmacognosy and Phytochemistry, Free University Berlin, Königin-Luise-Strasse 2-4, D-1000 Berlin 33, FRG
Schlegelmich F	FHN Krefeld, Frankenring, D-4150 Krefeld, FRG
Schmaus G	Institute of Pharmaceutical Biology, University of Würzburg, Mittlerer Dallenbergweg 64, D-8700 Würzburg, FRG
Then M	see Verzár-Petri
Verzár-Petri G	Institute of Pharmacognosy, Semmelweis Medical University, Üllöi út 26/III, H-1085 Budapest VIII, Hungary
Vita-Finzi P	Institute of Organic Chemistry, University of Pavia, Via Taramelli 10, I-27100 Pavia, Italy
Vostrowsky O	see Michaelis
Vuorela H	see Hiltunen
Weinstein V	Department of Organic Chemistry, Laboratory of Natural Products, Hebrew University, Jerusalem, 91904 Israel
Ziegler E	Research Laboratories, Aroma-Chemie, D-8551 Aufsess FRG

QUANTITATIVE ASPECTS OF FLAVOUR ANALYSIS BY EQUILIBRIUM AND DYNAMIC HEADSPACE GAS CHROMATOGRAPHY WITH CAPILLARY COLUMNS

B. KOLB

ABSTRACT

Equilibrium (E-HSGC) and dynamic (D-HSGC) headspace sampling procedures with capillary columns are compared. Split injection is widely used in headspace gas chromatography (HSGC) since a homogeneous gas mixture is injected and non linear split behaviour is unknown here. Splitless injection is recommended if headspace sampling is combined with cold trapping in order not to disturb the equilibrium, since a large sample is withdrawn over a longer period. In this case the well established methods for quantitave analyses with E-HSGC can be applied. These methods, including sample identification by pattern recognition, calibration by the technique of standard addition and the procedure of multiple headspace extraction (MHE) are discussed. Problems with D-HSGC techniques, particularly concerning an exhaustive stripping of all the volatile constituents from the sample without loss of the highly volatile ones by breakthrough in the adsorption tube are discussed also. In cases of mixtures with a wide range of volatilities, where the sample transfer can hardly be quantitative, the application of the MHE principle is suggested as a possible solution.

INTRODUCTION

Headspace gas chromatography (HSGC) is straightforward when volatiles have to be separated from a solid or liquid prior to the GC analysis, e.g. in cases where the sample itself cannot be injected. The analysis of the headspace vapours above foodstuffs by GC has been widely applied in aroma research, but various techniques have been used which are all called headspace techniques. The different meaning of what is called headspace may cause some confusion, particularly if the quantitative aspects are concerned.

In agreement with Wyllie et al. (1) there are good reasons to define headspace as the gaseous mixture surrounding a sample within a closed system, which according to all thermodynamic rules is in equilibrium. Contrary to this equilibrium headspace method there are a variety of methods which are used to isolate the volatiles from a sample by stripping with an inert gas. Such continuous stripping procedures need an intermediate storage zone, either a cold trap or an adsorption tube, in

A. Baerheim Svendsen and J.J.C. Scheffer (eds), Essential oils and aromatic plants, ISBN-13: 978-94-009-5137-2

which the volatiles are first separated from the stripping gas and further concentrated by band focusing, particularly necessary for the rigorous demands of capillary GC. Both the equilibrium and the dynamic headspace procedures have their particular advantages and typical applications, which should be discussed with special emphasis on the quantitative aspects. One of the most attractive properties of the equilibrium headspace procedure is its inherent simplicity, which lends itself to automation, while the expenditure to automate a dynamic headspace procedure is much more costly. The need for automation is not so well recognized, particularly not in research institutes where most of the work is related mainly to qualitative investigations. Most papers therefore deal with identification of flavour compounds preferably with GC-MS combination, and for such application neither the quantitative aspects nor a high sample throughput seem to be important. The situation, however, might be quite different when quality control in the production, e.g. food processing, is concerned. Differences in the flavour, which might be quite obvious to the human nose, are often only related to small differences in chromatographic peak relations. The significance of small differences in an analytical result, however, needs to be confirmed by statistical evaluation, and even only two samples require a series of chromatograms to be processed, and this immediately calls for automation. Automation in general improves the precision of the measurements and is thus a necessary prerequisite for statistical data processing and for the confirmation of the final result.

INSTRUMENTATION OF EQUILIBRIUM HEADSPACE GAS CHROMATOGRAPHY (E-HSGC)

HSGC is in fact a gas analysis and the same sampling techniques are often used as common in gas analysis. Gas-tight syringes or gas sampling valves with sample loops are thus widely employed. A gas syringe is the most popular device for transferring a gas sample from the headspace vial into a gas chromatograph. Apart from the contamination and adsorption problems with such a device, there is a more serious drawback, which is caused by the fact that a syringe is an open system during the sample transfer. Any pressure which has been generated by the increase of the partial vapour pressures of all the volatiles in the sample by heating the vial, also extends into the syringe. During sample transfer, however, the syringe needle is open to atmosphere and an undefined amount of the gas sample will be lost by expansion of the headspace gas through the needle. Even small differences in the humidity of two otherwise identical samples can simulate a big quantitative difference in the concentration of a volatile compound, although the same sample volume is withdrawn from the vial. Obviously not only the volume but also the pressure must be held constant to get the same sample amount onto the GC column, and independent of the sample composition. Such constant pressure sampling is most easily achieved if the vials are first pressurized with inert gas up to a constant pressure. By this procedure the sample amount becomes independent of the sample composition and good reproducibility of sampling is obtained (2). Several such systems are commercially available, from which the sampling principle as used in the HS-100 Automatic Headspace Sampler from Perkin-Elmer (Fig. 1) will be discussed. The whole instrument with all its possibilities is described elsewhere (2).

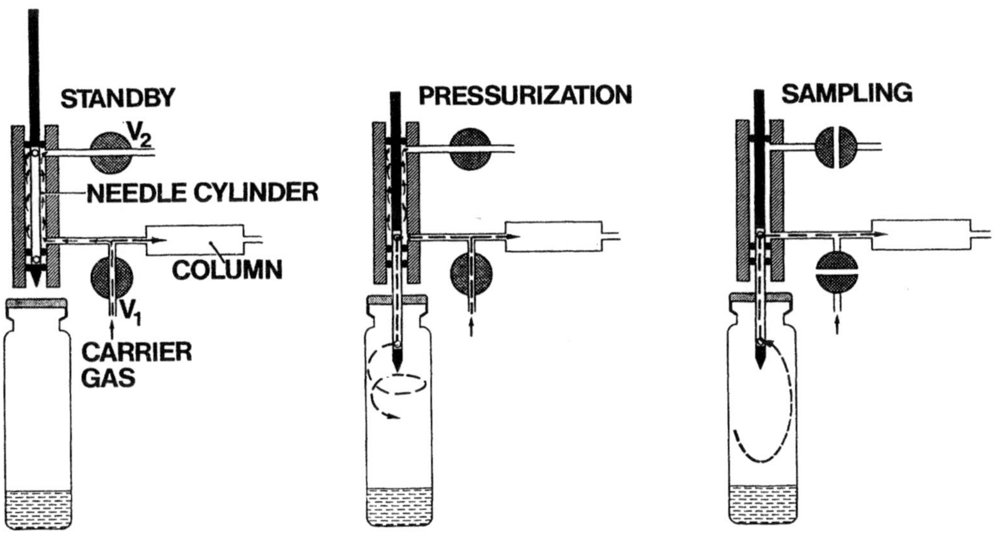

Fig. 1. Balanced pressure sampling with the Perkin-Elmer HS-100 Automatic Headspace Sampler.

The sampling device comprises a heated movable sampling needle with two vents and a solenoid valve, V1, in the carrier gas supply line. This needle moves up and down in a heated cylinder and is sealed by three O-rings. In the standby position the lower needle vent is placed between the two lower O-rings and thus sealed against atmosphere, while the carrier gas flows through valve V1 to the column. A small cross-flow purges the cylinder and is vented via valve V2. At the end of the selected thermostatting period, the sampling needle descends, pierces the septum cap and enters the headspace of the vial (pressurization). Carrier gas flows into the vial and pressurizes it either up to the head pressure of the column as shown in Fig. 1 or to any other higher pressure value, which is provided by an additional pressure regulator (not shown in Fig. 1) and a solenoid valve. The latter arrangement is recommended for columns with a small pressure drop such as wide bore capillary columns. At elapse of the preselected injection time (a few seconds) both valves open again. Carrier gas streams to the column and branches in order to pressurize the sample vial again. This immediately stops the injection and the residual sample vapours in the needle are flushed back into the vial, while the sample moves through the GC column.

This 'balanced pressure sampling principle' will now be discussed for its application with capillary GC. A small aliquot of the headspace gas is transferred onto the capillary column and the sensitivity is thus determined by this gas volume and the concentration of the volatile compounds in the headspace gas. The admissible gas volume of a capillary column is surprisingly small as compared to a packed column, due to the smaller cross-section. The headspace sensitivity of a packed column is therefore superior, while that of a capillary column is limited by the requirement to maintain the separation efficiency of the capillary, which needs a small starting bandwith of the sample plug. The admissible sample volume of a capillary column can easily be determined from the geometric dimensions of the capillary and the linear carrier gas flowrate in the column. If the starting bandwith of the volatile sample constituents should not exceed for instance 1 second, a sample volume of only 24μl is thus admissible with a capillary of 0.32 mm i.d. and a linear carrier gas flow of 30 cm/s. This is not much, but a larger sample volume will cause an unwanted band broadening.

These considerations apply both for split and splitless injection. But in gas analysis it does not make much difference which of these two injection techniques is used. The use of an inlet splitter needs a larger sample volume to be injected according to the split ratio, but again the starting bandwidth determine the actual sample volume in the column. Splitless injection, therefore, does not necessarily increase the sensitivity as in the case of liquid sampling, since a comparable solvent effect can not be used in gas sampling in general.

Split injection has no disadvantage and is widely used in HSGC. A non linear split behaviour is unknown here, since a homogeneous gas mixture is already withdrawn from the vial. On the other hand is splitless injection recommended if a small aliquot of the gas phase only should be taken from the vial in order not to disturb the partition system, or if enrichment techniques are applied, e.g. by using a cold zone, as discussed below.

A band focusing effect can be achieved even in HSGC by applying a cold zone between the sampling device and the column, as has been shown by Kuck (3), who cooled the first part of a glass capillary column by a stream of cold air. The same effect is obtained when the whole oven, including the column, is cooled down and even small temperature differences may be effective to get a remarkable band focusing, as demonstrated by comparing the chromatograms in Fig. 2. In chromatogram A the sampling time, which determines here the starting bandwith, was chosen to be short enough so as to get sharp peaks and good separation of the highly volatile compounds acetaldehyde and ethanol, while the lower volatile free fatty acids are eluted by temperature programming. A five times longer injection time deteriorates the early eluted peaks by a corresponding band broadening, while the peak shape of the free fatty acids is not affected at all. The peak heights, however, and thus the sensitivity, is increased about five times. In order to get similar good separation of the early eluted peaks too, the initial temperature had to be lowered further down and such an example of cryofocusing is shown in Fig. 3.

By using the possibility of such a cold zone, an enrichment by a factor 30-50 can be achieved, as has been shown by Kuck (3). It should be noted that in combination with splitless injection this enrichment effect is obtained without disturbing the partition equilibrium during the sampling time, since even with an increased sample volume the sampling time is short enough, compared with the slow diffusion of the volatiles between both phases, to prevent any change in the composition of the gas phase during the sampling time. It still remains an equilibrium headspace procedure and is not a dynamic technique. These are on the other hand all characterized by the fact that equilibrium is and should continuously be disturbed.

The combined splitless/cryofocusing capillary injection technique, therefore, allows all calibration methods for quantitative headspace analyses which are common in E-HSGC to be applied here also. These methods should, therefore, be discussed shortly before proceeding to the dynamic

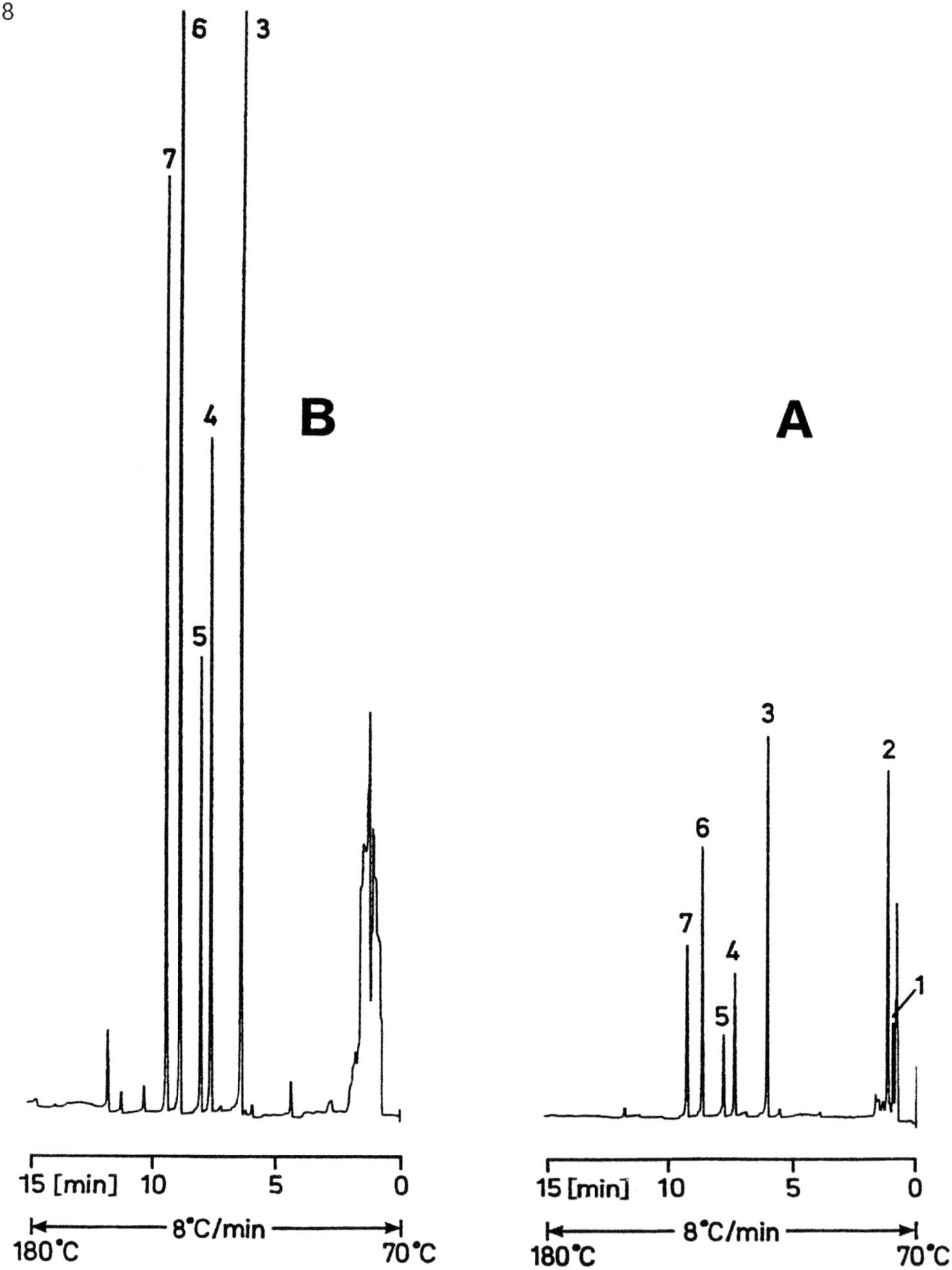

Fig. 2. Headspace analysis of flavour compounds from cheese. Perkin-Elmer Sigma 2000, HS-100; column, fused silica (25 m x 0.32 mm) coated with FFAP; balanced pressure sampling, splitless; sample, 2 g cheese, grinded, thermostatted 60 min at 90°C; chromatogram A, sampling time 4.8 s; chromatogram B, sampling time 24 s. 1 = acetaldehyde; 2 = ethanol; 3 = acetic acid; 4 = propionic acid (130 ppm); 5 = isobutyric acid; 6 = butyric acid; 7 = isovaleric acid (85 ppm).

headspace techniques.

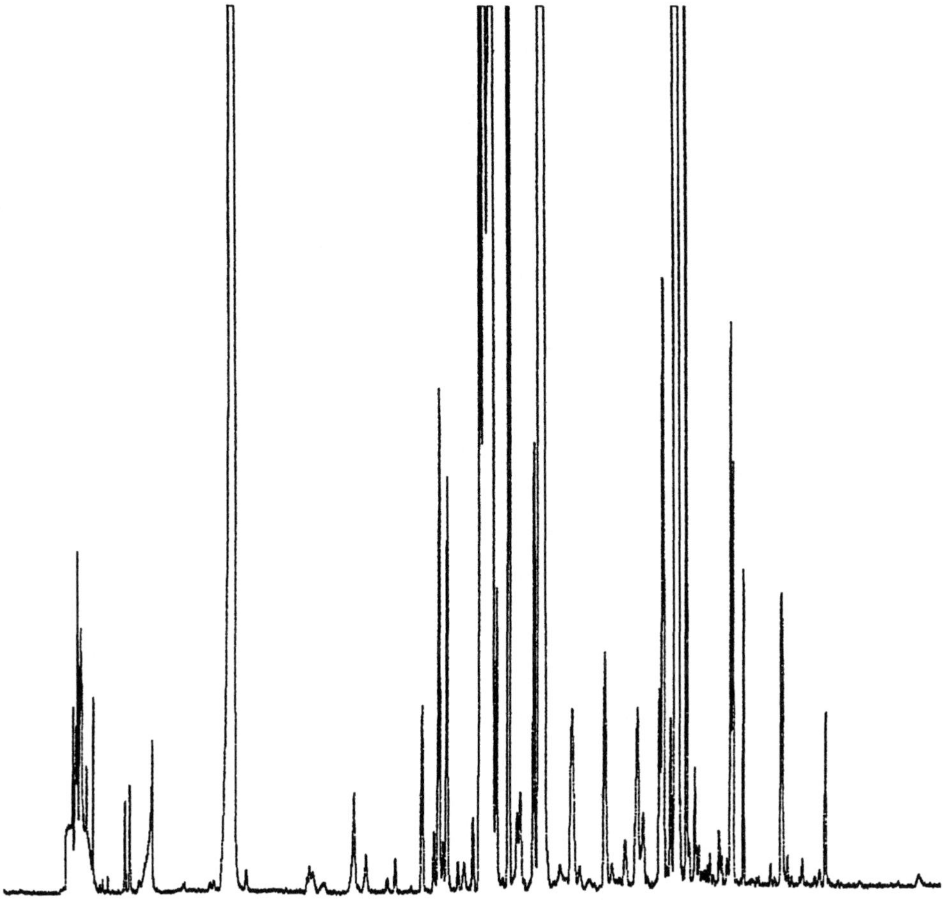

Fig. 3. Headspace analysis of tea aroma by cryofocusing. Perkin-Elmer Sigma 2000, HS-100; column, fused silica (50 m x 0.32 mm) coated with SE-54; temperature program, 60°C (6 min), then 60-140°C (6°C/min), after 4 min at 140°C, 140-200°C (20°C/min), then 200°C (25 min); carrier gas, Helium; sampling, splitless injection, 0.5 min; cryofocusing, -70°C, 200°C; sample, 400 mg tea (Earl Grey), 15 min at 80°C.

QUANTITATIVE EQUILIBRIUM HEADSPACE ANALYSIS

The first step in the development of a quantitative procedure for E-HSGC is the determination of both the time and the temperature which are necessary to obtain a state of equilibrium. This is easily carried out by processing a series of samples from the same material with increasing thermostatting time at a preselected temperature until the peak size remains constant.

To begin with a detailed discussion of the quantitative aspects, it seems to be necessary to mention the use of internal standards, which are often added to a sample if the reproducibility of sampling is not satisfactory, as is often found with manually operated gas syringes. With automatically operated pneumatic sampling systems, however, the instrumental precision favours the application of external standards rather than the addition of internal standards with their necessarily high volatility, which are often very cumbersome to mix precisely to each sample.

Quantitative Analysis of the Aroma Composition

When a state of equilibrium is achieved, the phase distribution of a volatile solute is determined by its partition coefficient K according to Equation 1, and the peak size A in the resulting headspace chromatogram is thus proportional to the gas phase concentration Cg (Equation 2), while Cl is the concentration of the liquid sample.

$$\text{Equation 1:} \quad K = Cl/Cg \qquad \text{Equation 2:} \quad A \sim Cg = Cl/K$$

The chromatogram in fact shows the actual composition of the gas phase above a sample, and a quantitative analysis of the aroma composition needs only the detector response factor of each compound to be included in the calculation, as usual in GC analysis. The result, however, is the aroma composition in a closed system, while for an open environment the MHE procedure, which will be discussed later, may be more suitable in order to follow the change in the aroma composition with the time.

Sample Identification by Pattern Recognition

Without any calibration a headspace chromatogram may be used as a fingerprint to characterize or identify a sample. Again, an equilibrium provides for the required reproducibility. This application, however, is preferably carried out on the basis of a list of identified compounds and their quantitative peak area relations, rather than using the analogue chromatogram itself. The simple comparison of two chromatograms may often be quite informative for an experienced analyst; with increasing numbers of reference chromatograms, however, this application can be carried out only by pattern recognition with a computer. This approach requires that the high number of data be independent from chromatographic and instrumental parameters, as well as from the influence of the column stability.

Quantitative Analysis of Liquid Samples

Most headspace applications ask for concentration of a volatile compound in the liquid or solid sample, rather than in the gas phase above it. This is possible too, and the analysis of a liquid sample by E-HSGC is discussed first, since it is the most simple example. The equilibrium is determined by the partition coefficient K according to Equation 1, and the peak size A in the headspace chromatogram is proportional to the gas phase concentration Cg, as described by Equation 2. The relationship of Equation 2 needs calibration, which now includes not only the detector response but also the partition coefficient K and thus the influence of the liquid sample matrix on the volatility.

If the pure sample matrix is available, there is no problem in preparing a calibration standard, but such samples are rare in HSGC. Sometimes the matrix can be simulated, e.g. for flavour analysis of an alcoholic beverage the matrix is prepared from an aqueous ethanolic solution. Most practical samples for HSGC, however, have an undefined matrix, which is included in the calibration procedure at best by the technique of standard addition as shown in the example of Fig. 4.

The technique of standard addition is the most universal calibration method in HSGC, but requires two analyses to be carried out from each sample, one from the original sample and another after addition of a certain amount of the compound, the concentration of which is calculated from the difference in the peak size between both chromatograms. Obviously any matrix effects are compensated. It is good practice in HSGC to start with this procedure if a new and unknown sample type has to be analyzed. This procedure, however, is restricted to liquid samples and cannot be applied to solids unless the solid material is first dissolved and the resulting solution used for HSGC. This solution approach is widely applied for the determination of monomers in polymers, but suffers the disadvantage of decreased sensitivity due to the additional dilution and increased solubility of the volatiles in the solvent.

Quantitative Analysis of Solid Samples by the Technique of Multiple Headspace Extraction (MHE)

The problem with solid samples for the quantitative analysis of a volatile in it is a practical one, because it is clearly impossible to add trace concentrations of volatile compounds homogeneously to such samples for applying the technique of standard addition. In principle, however,

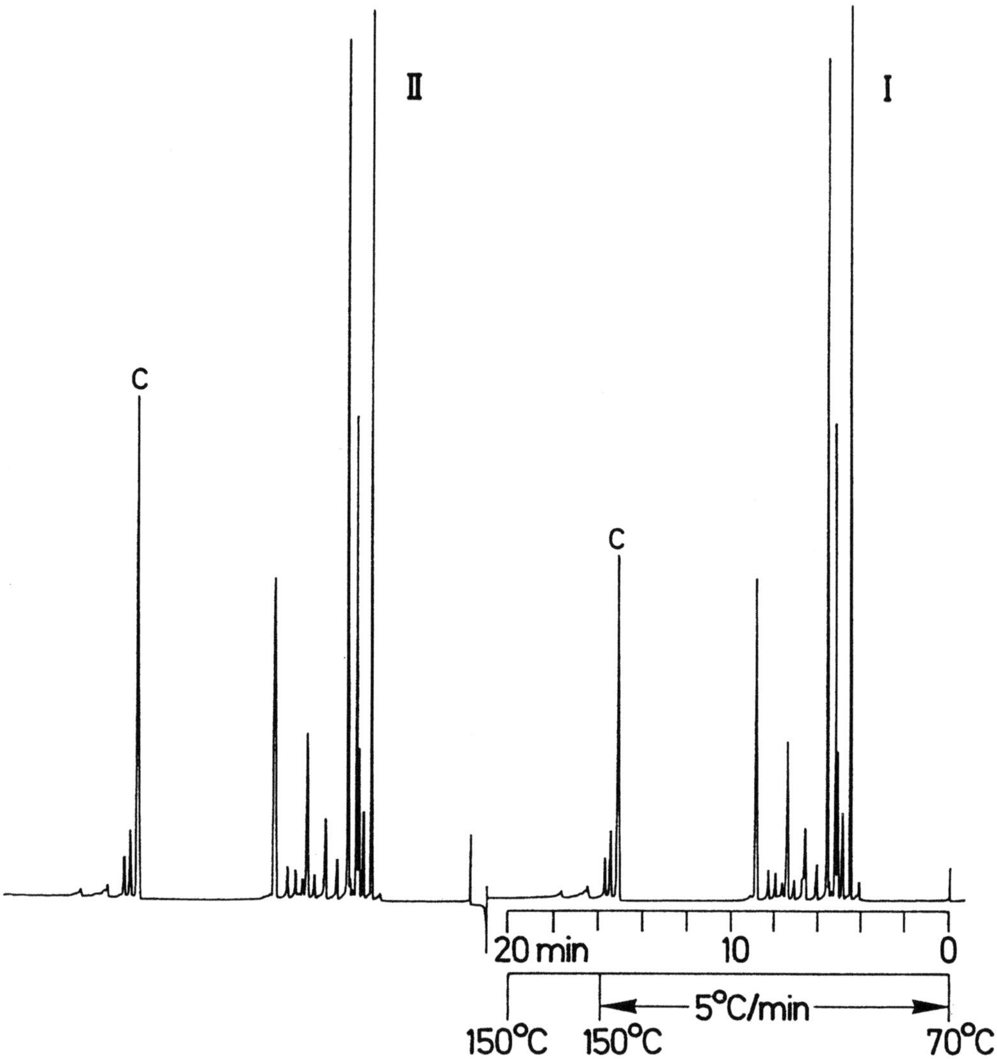

Fig. 4. Quantitative headspace analysis of 1.1% camphor (C) in massage cream by the technique of standard addition. Perkin-Elmer Sigma 2, HS-6; column, glass capillary, coated with Carbowax 20M; split injection; chromatogram I, 1 g sample at 80°C; chromatogram II, 1 g sample + 5 mg camphor.

the same result can be achieved if a certain fraction of the volatile solute is taken away from the sample rather than added. With the pneumatic sampling device, as described previously, the headspace vials remain pressurized after each injection and the easiest way to remove a part of the volatile solute is by venting the pressurized vial to atmosphere, e.g. by piercing the septum of the vial manually with a syringe needle. With the HS-100 Automatic Headspace Sampler this venting procedure as well as its instrumental details are described elsewhere (2-6). If, for example, the pressure in the vial is one bar, half of the gas phase is vented and the equilibrated partition system will thus be disturbed seriously. After the necessary equilibration time, however, a second headspace analysis will result in a chromatogram in which the peaks are smaller. If repeated several times, all the volatiles are stripped off and a continuous stripping process is thus simulated by a stepwise approach. If carried out until or close to exhaustion, a series of chromatograms is obtained, from which the peak areas for each volatile must be summed to get an area total. This total is proportional to the total amount of that compound present in the headspace vial in both phases and not just in the gas phase. This area total is thus indpendent from the distribution ratio of that volatile between both phases, thus eliminating any influence of the sample matrix. This area total, however, must not necessarily be determined experimentally, but can be calculated by mathematical extrapolation, since the stepwise extraction at equal time intervals preceeds according to an exponential function. The area total ΣA is derived as the sum of a geometric progression. Only a few of these determinations are needed to apply linear regression calculation. In practice two determinations have often been found sufficient to calculate this area total ΣA from the first (A_1) and the second (A_2) area value from the two successive chromatograms according to the following equation

Equation 3: $\Sigma A = A_1^2 / (A_1 - A_2)$

Since this simple two-step procedure needs two chromatograms to be run on the sample, it is comparable to the technique of standard addition.

The area total, obtained either by linear regression calculation or by the simpler two-step procedure, needs to be calibrated like any peak area in every GC analysis, but this calculation involves the detector response and the instrumental conditions, but no longer the matrix effects. Such a calibration standard can thus be prepared from a sample with either a similar or a different matrix and even the pure vapour of that volatile compound in an empty headspace vial can be used. Such a vapour sample can easily be prepared if a small amount - a few microliters - of the respective compound or a dilute solution is vapourized in an empty vial.

The MHE procedure can be applied for solid samples if these are present either as a fine porous powder or as a thin film; it can be used with liquid samples also, but requires a high volatility of the sample constituents, e.g. partition coefficient smaller than 100. Otherwise, the difference in the peak sizes between two successive chromatograms will be too small as to allow extrapolation for exhaustive extraction. Liquid samples which do not meet this requirement are preferably analyzed by the technique of standard addition, as discussed above.

As a practical example, Fig. 5 shows the first three chromatograms of a five-step MHE procedure for the determination of 73 µg/g trans-1,2-dichloroethylene in coffee powder.

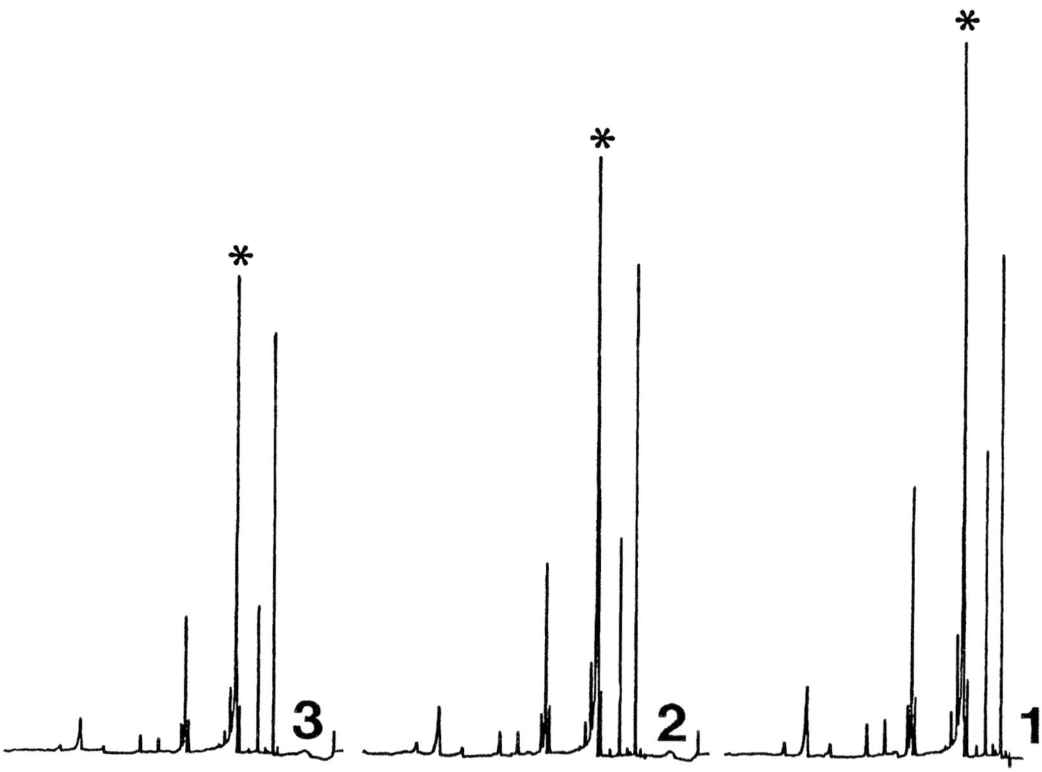

Fig. 5. Determination of 73 ppm (w/w) trans-1,2-dichloroethylene in coffee powder by multiple headspace extraction (MHE). Perkin-Elmer Sigma 2000, HS-100; column, fused silica (50 m x 0.32 mm) coated with SE-54; 70°C isothermal; carrier gas, Helium; split injection; sample, 474 mg coffee + 100 µl water (displacer), 30 min at 80°C.

The plot in Fig. 6 shows the determination of this result by linear regression calculation; the area values are plotted in a semilogarithmic scale versus the number of successive analyses and a linear relationship is thus obtained, which enables the extrapolation of the area total (final area) in both the coffee sample and the calibration standard. The latter was prepared by injecting 3 µl of a solution of 5.06 mg/ml trans-1,2-dichloroethylene in dioxan into an empty headspace vial. This calibration standard was analyzed three times with the MHE procedure, and the resulting peak areas are listed in the plot as shown in Fig. 6, while the chromatograms are not shown here.

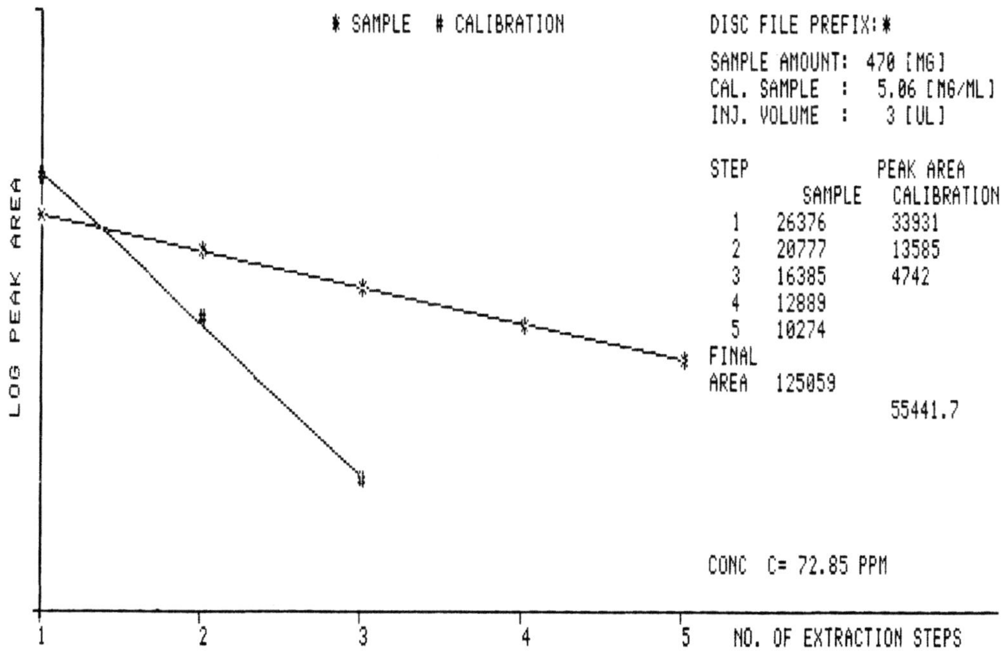

Fig. 6. Quantitative determination of 73 ppm (w/w) trans-1,2-dichloroethylene in coffee powder by a five-step MHE analysis (see Fig. 5) with linear regression calculation.

INSTRUMENTATION FOR DYNAMIC HEADSPACE GAS CHROMATOGRAPHY (D-HSGC)

There are numerous descriptions of various devices for performing what is termed as dynamic headspace sampling. Since all these procedures are based on stripping the volatiles from a sample by a continuous flow of inert gas, an intermediate storage zone is essential because the stripping process is in general too long to be performed on-line with the gas chromatograph. Additional focusing of the original broad concentration profile becomes necessary, and at the same time the concentration in the gas phase must be increased, since depending on the volatility of the respective compounds, a highly diluted gas sample is obtained. Such a storage zone helps to separate the volatiles from the bulk of inert stripping gas, and in this way a remarkable enrichment is achieved. This is the main advantage of the dynamic versus the equilibrium headspace procedure.

The separation of the volatiles from the stripping gas can be performed either by absorption in a solvent or by adsorption on a solid adsorbent. If a strong adsorbent is used, such as charcoal, silica gel or alumina, the subsequent desorption needs in general to be performed by displacement with a solvent and a dilute solution is thus obtained, as in the case of absorption in a liquid solvent. This, on the other hand, reduces the enrichment effect, since again an aliquot of the dilute solution can only be injected by a syringe. For highly volatile compounds the resulting solution alternatively can be analyzed by E-HSGC instead of the usual syringe injection. In this case a low volatility solvent, such as benzyl alcohol is preferably used as a displacer (7-9). Benzyl alcohol is eluted late in the resulting headspace chromatogram or may be removed by backflushing (8, 9).

Thermal desorption is more popular since the enrichment effect is enhanced, but needs weak adsorbents, such as porous polymers, from which Tenax GC has found widespread application. Thermal desorption from an adsorption tube, however, is not fast enough to be used directly for sample introduction into a capillary column, and again an additional focusing zone is necessary from which a rapid second stage desorption is started. One of the useful properties of porous polymers as adsorbents is the low retention of water, which has a retention index of only about 300 on such materials and it is, therefore, possible to remove most of the water by adjusting the temperature of the adsorption tube such that water breaks through, while the volatiles of interest are retained. In this case, the second focusing zone can be a cold trap, e.g. by cooling the first part of a capillary column according to the procedure as described by Kuck (3) or by using again a second trap filled with porous polymers. For this purpose the trap can be designed much smaller, since the bulk of the stripping gas has been removed already by the first adsorption tube. Various combinations are thus possible. If the capillary column is used as a trap by cooling either its first part or the whole column, the cooling temperature should be at least -80^{0}C if high volatile compounds have to be trapped efficiently (10). If for the second focusing zone adsorption on porous polymers is used, it may be combined with additional cooling, which in this case needs to be less rigorous.

A typical instrumental set-up including the necessary flow pattern is shown in Fig. 7. A detailed description is given elsewhere (11, 12).

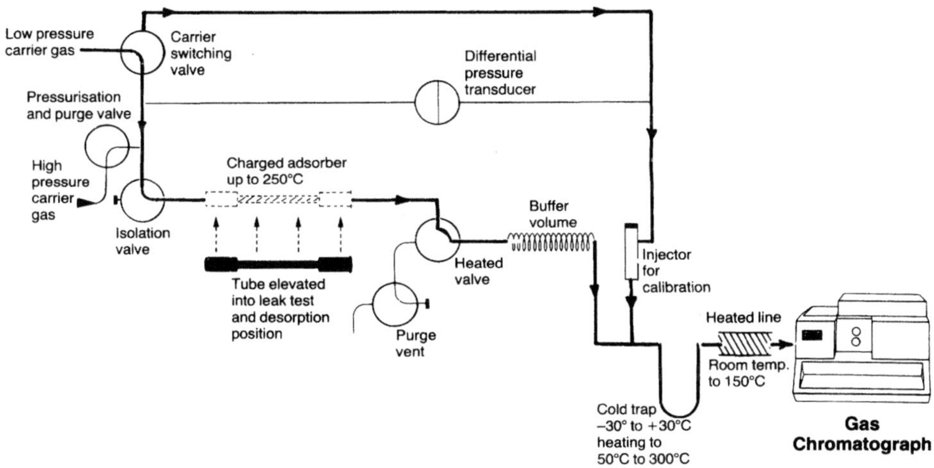

Fig. 7. The desorption process with an automatic thermal desorption system Perkin-Elmer ATD-50.

The ATD-50 system can be attached to any gas chromatograph and contains a horizontal circular table with a capacity of 50 adsorption tubes. A charged tube is automatically transferred into the desorption oven, connected into the flow system, purged, pressurized and leak tested. It is then heated to the desorption temperature and the desorbed compounds are swept onto an electrically cooled trap, from which second stage desorption is induced by ohmic heating with a heating rate of 1000^{0}C/min to a controlled temperature to allow a narrow band of desorbed sample to be swept onto the capillary column. The ATD-50, however, does not include the stripping device, because this instrument is mainly used for personal air monitoring, where the tubes are preferably charged by diffuse sampling, although pumping is possible as well. The chromatogram in Fig. 8 shows such an application for diffusive sampling of a perfume headspace, by holding a Tenax GC packed ATD-50 tube over a bottle of perfume for 15 min.

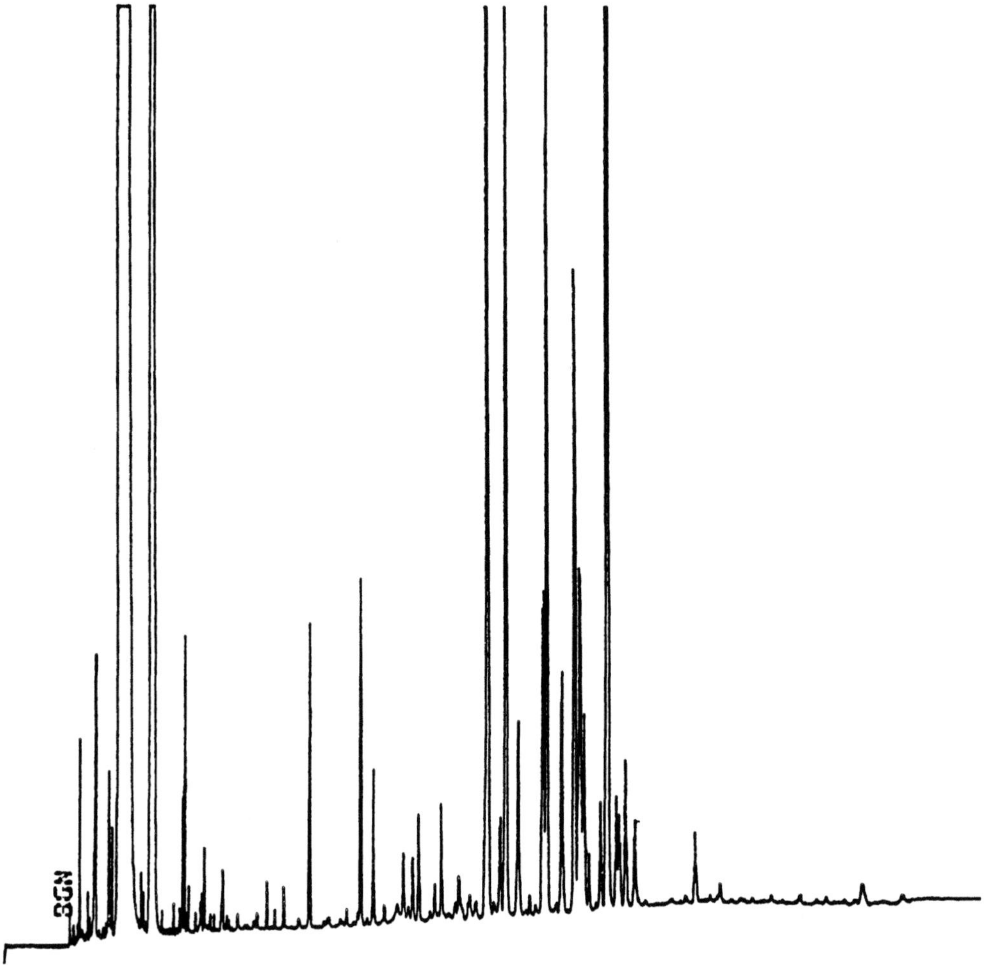

Fig. 8. Analysis of perfume headspace by adsorption on Tenax GC. Perkin-Elmer ATD-50; column, fused silica capillary coated with Permaphase DMS (dimethyl silicon); temperature program, 50-200°C (4°C/min); sampling: 15 min passive sampling with Tenax GC packed ATD-50 tube over a perfume bottle.

QUANTITATIVE ASPECTS OF DYNAMIC HEADSPACE GC

D-HSGC has been used so far for qualitative purposes, mainly as an enrichment technique to get sufficient amounts of compounds for further identification, preferably by GC-MS. Little attention has been paid to quantitative applications, but with increasing applications of HSGC for precise quantitative purposes, these aspects might become more interesting in the future.

The general problems of stripping/adsorption is breakthrough of the highly volatiles and complete stripping of the low volatility compounds from the sample on the other side, particularly if samples with a wide range of volatilities have to be investigated. Quantitative determination by D-HSGC needs the following conditions to be fullfilled:

- the stripping yield of all the volatiles must be complete;
- the adsorption of all the volatiles must be complete, without breakthrough of the highly volatile compounds.

Only if these requirements are fullfilled, it is possible to determine the concentration of each compound by standard GC methods. If the stripping yield is not complete, the degree of extraction must be included in the quantitative calculation and this requirement can hardly be realized, since again the influence of the sample matrix and small variations in the matrix composition are involved, as already discussed above for E-HSGC. This general problem of the dynamic headspace procedure should be discussed shortly using as examples a few typical compounds of different polarity and hence widely differing partition coefficients in aqueous solution, see Table I.

Table I. Partition coefficient K in aqueous solution at 70°C and retention volume RV in liters per g Tenax GC at 20°C

Compound	Boiling point in °C	K at 70°C	RV in liters at 20°C
Ethanol	78	300	2
Methyl ethyl ketone	80	40	33
Dichloromethane	40	2.3	2
Tetrachloroethylene	121	0.2	485

If the compounds such as listed in Table I have to be stripped off completely from an aqueous solution and adsorbed in a Tenax GC trap, it is obvious from the figures in the Table that a large volume of stripping gas is necessary to extract an alcohol from an aqueous solution, while on the other hand a long adsorption tube has to be used to prevent breakthrough due to the small retention on Tenax GC. Non polar compounds such as the halogenated hydrocarbons are much easier to handle, since a high volatility from aqueous solution is combined with a strong retention on the adsorbent.

It is therefore not only the volatility of a compound - as expressed by its boiling point - which describes its behaviour in dynamic headspace sampling, but also the matrix of the sample - as expressed by the partition coefficient and the retention properties on the adsorbent. Alle these factors have to be taken into account if the conditions of such a dynamic headspace procedure have to be established. A high retention in the sample matrix does not necessarily imply a similar retention on the adsorbent, and both effects may influence the volatility of a compound by several orders of magnitude. All combinations are possible and found in practice. If volatile compounds with a wide range of volatilities and polarities have to be analyzed by dynamic headspace techniques in a flavour analysis, a quantitative transfer of the headspace sample can hardly be realized. In this situation, however, it should be possible to apply the principle of the MHE procedure to the dynamic headspace techniques also, because there is no reason why it should be restricted to E-HSGC. In both cases an incomplete gas extraction is controlled by several repeated analyses, and the quantitative result derived by mathematical calculation. In fact, the analyst is interested in the information, rather than in the total physical amount of a substance, contrary to a preparative working chemist, and a quantitative extraction yield is not important if the final result can be obtained by calculation based on an incomplete extraction.

In practice the MHE procedure needs at least two repeated analyses to be carried out of the sample, and from the difference in the peak sizes of each compound the original amount in the sample can be determined by extrapolation, as discussed above.

CONCLUSIONS

Both equilibrium (E-HSGC) and dynamic (D-HSGC) headspace sampling techniques are compared. E-HSGC is easier to use for quantitative determinations, and the available methods are discussed in detail. D-HSGC is on the other hand more sensitive, since it includes a remarkable enrichment of the volatile sample constituents, while its quantitative application is more difficult. For a wide range of volatilities the stripping yield is hardly quantitative for all compounds without breakthrough of the highly volatiles. As a possible solution to this dilemma the application of the multiple headspace extraction procedure (MHE) to the various dynamic headspace techniques is suggested. Such an approach, however, needs several repeated determinations from each sample to be carried out, and the time expenditure immediately calls for automation. However, it is nowadays generally accepted that good quantitative results in chromatography in fact need automated equipment.

REFERENCES

(1) Wyllie SG, Alves S, Filsoof M, Jennings WG. 1978, in: Analysis of Food and Beverages (Charalambous G, Ed.), New York, Academic Press Inc., p. 1.
(2) Closta W, Klemm H, Pospisil P, Riegger R, Siess G, Kolb B. 1983, Chromatography Newsletter 11:13, Perkin-Elmer.
(3) Kuck M. 1980, in: Applied Headspace Gas Chromatography (Kolb B, Ed.), London, Heyden & Son Ltd., p. 12.
(4) Kolb B, Auer M, Pospisil P. 1981, Applied Chromatography No. 35E, Ueberlingen, Bodenseewerk Perkin-Elmer & Co. GmbH.
(5) Kolb B. 1984, in: Topics in Forensic and Analytical Toxicology (Maes RAA, Ed.), Amsterdam, Elsevier, p. 119.
(6) Kolb B. 1984, in: Analysis of Food Contaminants (Gilbert J, Ed.), London and New York, Elsevier Applied Science Publishers Ltd., p. 117.
(7) Bencsath FA, Drysch K, List D, Weichardt H. 1978, Applied Chromatography No. 32E, Ueberlingen, Bodenseewerk Perkin-Elmer & Co. GmbH.
(8) Kolb B, Pospisil P. 1980, Applied Chromatography No. 33E, Ueberlingen, Bodenseewerk Perkin-Elmer & Co. GmbH.
(9) Kolb B, Pospisil P. 1980, Chromatography Newsletter 8:35, Perkin-Elmer.
(10) Pankow JF. 1983, J. High Resol. Chromatogr. & Chromatogr. Commun. 6:292.
(11) Deckert HG, Foelster U, Rueck A. 1984, LaborPraxis 8:130.
(12) Hurrell RA. 1981, International Environment & Safety, June: 18.

QUANTITATIVE HEADSPACE GAS CHROMATOGRAPHY IN THE ANALYSIS OF VOLATILE OILS AND AROMATIC PLANTS

R. HILTUNEN, H. VUORELA, I. LAAKSO

ABSTRACT

The peak area of components analyzed by headspace gas chromatography (HSGC) is determined by a number of parameters, such as equilibration period and temperature, the vapour pressure of the components, etc. In order to optimize quantitative HSGC determinations, these parameters must be taken into account. In the present study the effects of these parameters as well as the effects of varying calibration methods were investigated.

Various herbs and the volatile oils isolated from them by hydrodistillation were used for the investigation. Analyses were carried out on a gas chromatograph DANI HR 3800 equipped with glass capillary columns and connected with a headspace sampler DANI HSS 3850. Comparison of equilibration periods (0.5, 2, 5, 10, 20 and 45 min) and temperatures (60, 90, 120 and 150°C) indicated that an equilibration period of 15-20 min yielded optimum results for liquid volatile oil samples at the temperatures used. Solid samples such as herbs, however, required a longer equilibration period. The reproducibility of the HSGC analyses was comparable or better than that of classical GC analyses.

Percentage normalisation, calibration with internal or external standards, standard addition, and the multiple headspace extraction (MHE) technique were used in the quantitative determinations. All calibrations gave the same results within the acceptable analysis errors. The most reliable methods proved to be the standard addition to the matrix and the MHE methods.

INTRODUCTION

Gas chromatographic (GC) methods in which the volatile components in the gas space around liquid or solid samples are determined indirectly, have been used since the early days of GC. In the 1960's such methods were known under the general name of 'vapour phase equilibrium', and later on as 'headspace analysis method' and 'headspace GC method'. In simplified form, the methods involve analyses in which the components of the sample are allowed to reach a state of equilibrium — determined by their specific equilibrium constants — between the gas and solid or liquid phase. A sample from the equilibrated gas phase is transferred either manually or automatically to the gas chromatograph for analysis. The gas sample is either introduced directly into the GC column, or concentrated using liquid or solids traps

A. Baerheim Svendsen and J.J.C. Scheffer (eds), Essential oils and aromatic plants, ISBN-13: 978-94-009-5137-2

and then transferred into the GC column by means of thermal desorption or liquid extraction.

Headspace gas chromatography (HSGC) and its different methodological applications have recently been discussed in detail in a publication covering both the theoretical principles and practical procedures (1). During the past 20 years, HSGC methods have been widely applied in the fields of environmental analysis, clinical research, industrial processing and quality control as well as in microbiological and food chemistry studies. In a tremendous number of publications it is claimed that HSGC is a rapid, simple, precise, sensitive, reproducible and even non-destructive technique, which can easily be used for both qualitative and quantitative determination of volatile compounds present in liquid or solid materials. A large number of studies cover the use of HSGC in quantitative determinations. The publications discuss different sample handling techniques, calibration methods, matrix effects and the effects of equilibration period and temperature on the equilibrium state and on the sample itself (2-16).

Despite the fact that HSGC has been widely used in qualitative and quantitative studies on a wide range of organic compounds, there are only a few reports available concerning the use of HSGC in the analysis of volatile oils and aromatic plants. HSGC has been used in a number of studies for the quality control of aromatic herbs (17-20), mainly in 'fingerprint' analyses. The dynamic HSGC technique has been applied in the quantitative determination of some monoterpene hydrocarbons – MTHC – (21) and also in calculating the relative proportions of these hydrocarbons (20, 22). One study has been performed on HSGC in the quality control of flavours in soft drinks (23). The quantitative multiple headspace extraction (MHE) method, described by Kolb (5) and Kolb et al. (3, 9), has been used in studies on volatile oils and aromatic compounds (17, 23). According to the literature, MHE is suitable at least for the most volatile compounds (3, 9, 23).

The aim of the present study was to optimize the HSGC technique in order to obtain precise quantitative results for volatile oils and especially for MTHC, oxygenated monoterpenes and sesquiterpenes in liquid and solid samples. Special attention was paid to equilibration parameters such as time and temperature, and also to other parameters which affect the proportion of the components in the gas phase. In addition, a number of different calibration techniques were tested in analyses involving both liquid and solid matrices.

EXPERIMENTAL

Instrumentation

The headspace gas chromatographic (HSGC) analyses were carried out on a gas chromatograph DANI HR 3800 equipped with a headspace sampler DANI HSS 3850. Columns: glass (20 m x 0.35 mm i.d.) coated with OV-1 and Carbowax 20M respectively; oven temperature programmed, 55-200°C (5°C/min), or isothermal, 70°C (for MTHC) and 170°C (for other compounds); injection port, 250°C; detector, FID, 250°C; carrier gas, Nitrogen, 0.75 kp/cm^2; auxillary gas, Nitrogen, 0.70 kp/cm^2; split ratio, 1:20.

The sample vials (20 ml) were tightly sealed with a rubber septum covered with teflon. Nitrogen was used as the pressurisation gas (2.8 bar). The volume of the sample loop was 250 µl. The backflushing rate for the sample loop, sample probe and the sample transfer tube was 30 ml/min, and the backflushing delay was set at 9 s.

Equilibration periods of 0.5, 2, 5, 10, 20 and 45 min were used. The multiple headspace extraction (MHE) analyses of thyme herb, chamomile flowerheads and the volatile oil of chamomile were carried out using an equilibration period of 45 min prior to the first extraction step. In the case of chamomile the injections in the subsequent extraction steps were carried out at 20 min intervals, and in the case of thyme at 35 min intervals. Pressure equilibration was maintained automatically in the sample vials during injection. Oil bath temperatures of 60, 90, 120 and 150°C were tested. The transfer tube was kept at 150°C.

Material

Liquid samples:

- an artificial mixture of monoterpene hydrocarbons (MTHC) of the following proportions (by weight): α-pinene (purity, verified by GC, 96.2%) 20.5%, β-pinene (purity 83.8%) 19.3%, 3-carene (purity 90.0%) 12.4%, limonene (purity 97.0%) 27.6%, β-phellandrene (purity 52.2%) 18.8% and terpinolene (as an impurity) 3.4%; the mixture was diluted 1:20 with n-hexane; 2 µl doses were placed in the vials;
- α-bisabolol 1:20 in n-hexane, 2 µl per vial;
- thymol 1:20 in n-hexane, 2 µl per vial;
- hydrodistilled volatile oil of chamomile flowerheads (*Matricaria recutita*) 1:20 in n-hexane, 2 µl per vial;
- hydrodistilled volatile oil of thyme herb (*Thymus vulgaris*), 0.2-1 µl

per vial;

- a commercial chamomile preparation, 10 μl per vial.

Solid samples:

- powdered chamomile flowerheads, 10 mg per vial;
- powdered thyme herb, 20 mg per vial;
- powdered *Thymus serpyllum* herb, 30 mg per vial;
- sliced shoots of *Larix* spp. (*L. sibirica*, *L. decidua*, *L. leptolepis*, *L. leptolepis* x *sibirica* and *L. decidua* x *leptolepis*) 100 mg per vial;
- chopped flowerheads of *Tanacetum vulgare* (camphor, thujone, sabinene and umbellulone chemotypes), 30 mg per vial;
- chopped needles of *Pinus sylvestris* (high 3-carene and low 3-carene chemotypes) 100 mg per vial.

Water (2 μl) was added to all sample vials. The amount of n-hexane was kept constant at 2 μl per vial.

RESULTS AND DISCUSSION

In contrast to classical GC, the area of the peaks produced in HSGC is dependent on a number of different factors. These include equilibration temperature and period, sample size, and volume of the vial. However, the most important factor is the partial pressure of the compound being analyzed. In addition, the matrix of the sample also exerts an effect on the components being studied, and hence on the area of their peaks.

Effect of equilibration temperature

In general, increasing the temperature, while keeping the other parameters constant, causes an increase in the peak area of the individual compounds. However, the dependence on temperature is not linear. Neither is the trend the same for different compounds. Furthermore, the diffusion of the MTHC from the different matrices into the gas phase is also variable. It can be seen from Fig. 1 that the proportion of MTHC does not increase to a significant degree over the temperature range tested (60-150^{0}C), if the sample is in liquid form, i.e. an oil. However, if the sample is plant material, the proportion of MTHC changes drastically over the temperature range in question. Thymol is an exception: it behaves in a similar fashion in both solid and liquid matrices. The behaviour of α-bisabolol in the gas phase also differs from the above. The increase in the proportion of α-bisabolol is largest over the temperature range 60-90^{0}C. Increasing the

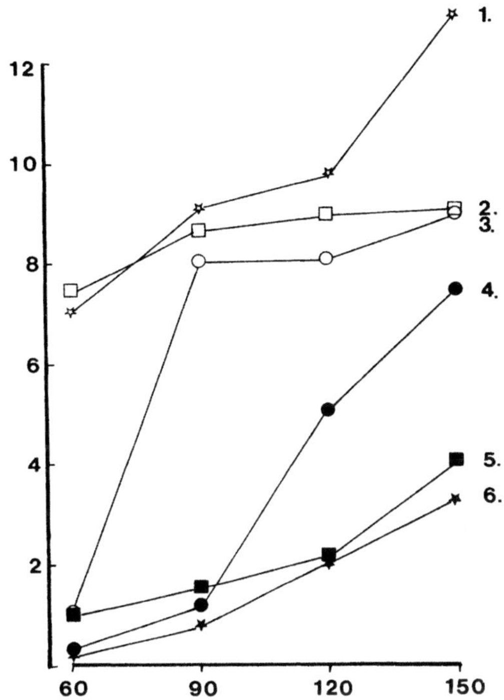

Fig. 1. The effect of bath temperature on the peak areas of some terpenes: 1. thymol; 2. mixture of MTHC; 3. α-bisabolol; 4. α-bisabolol in chamomile flowerheads; 5. MTHC in thyme herb; 6. thymol in thyme herb. Abscissa: bath temperature (^{o}C); ordinate: peak area x 10^4.

temperature further has little effect on the proportion of α-bisabolol.

It can also be seen from Fig. 1 that the increase of α-bisabolol is rapid over a narrow temperature range, increasing three-fold when the temperature is increased from 60°C to 65°C. For this reason, and since the temperature dependence of different compounds varies, it is of prime importance that the bath temperature is kept as constant as possible.

Effect of equilibration period

When the matrices are unknown, it is important to optimize the equilibration period. As the rate of equilibration of the different components between gas and liquid phase is different, the composition of the gas phase is constantly changing. So, if the analysis is carried out before the state of equilibrium is reached, spurious results will be obtained. On the other

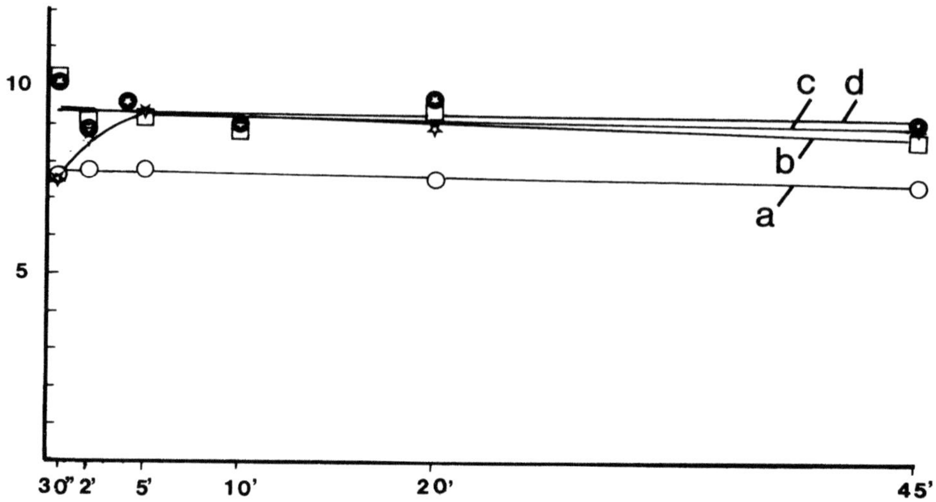

Fig. 2. The effect of equilibration period on the total peak area of the MTHC mixture in HSGC analyses at various temperatures: a = 60°C; b = 90°C; c = 120°C; d = 150°C. Abscissa: equilibration period; ordinate: peak area x 10^4.

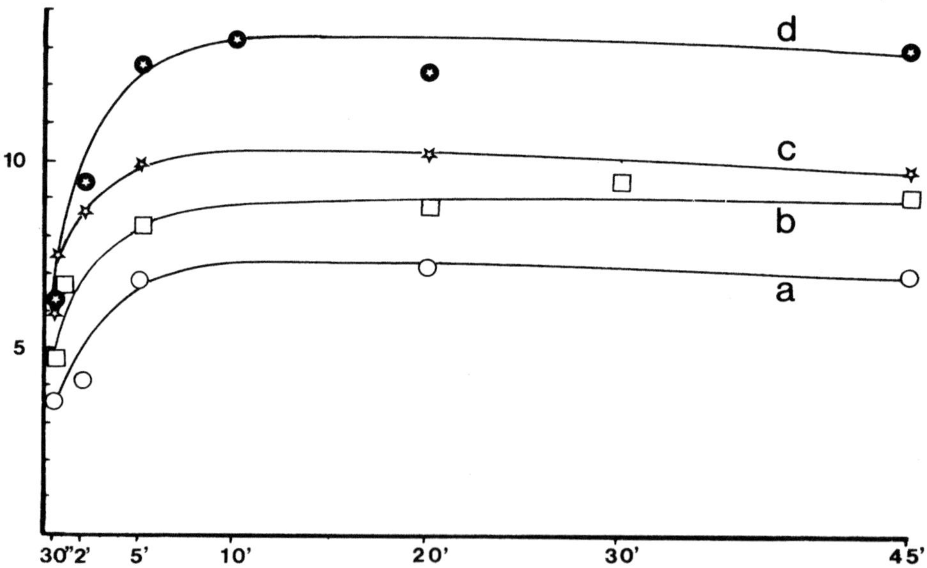

Fig. 3. The effect of equilibration period on the peak area of thymol in HSGC analyses at various temperatures: a = 60°C; b = 90°C; c = 120°C; d = 150°C. Abscissa: equilibration period; ordinate: peak area x 10^4.

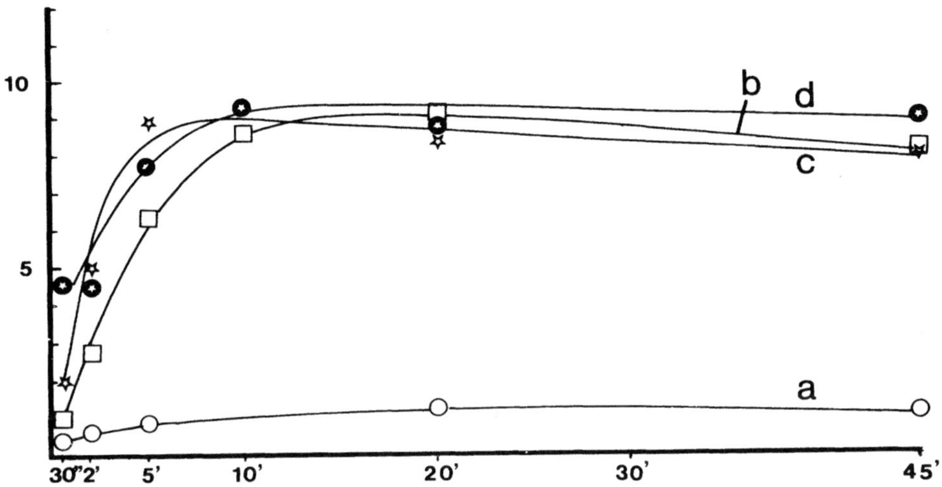

Fig. 4. The effect of equilibration period on the peak area of α-bisabol-ol in HSGC analyses at various temperatures: a = 60°C; b = 90°C; c = 120°C; d = 150°C. Abscissa: equilibration period; ordinate: peak area x 10^4.

hand, if the state of equilibrium is maintained for a long period, especially at high temperatures, decomposition, rearrangement and oxidation may occur. Leaks from the sample vial may also become significant.

When the equilibration periods for liquid matrices, i.e. the mixture of MTHC, thymol and α-bisabolol, were studied at four different temperatures, it became apparent that thymol and α-bisabolol did not attain equilibration until after at least 15-20 min (Figs. 2-4). The MTHC mixture, on the other hand, required only about 5 min. However, the precision was initially very low. The proportion of oxygenated compounds in the gas phase increased three- to four-fold during the first few minutes. A slight decrease in concentrations was observed when equilibration periods longer than 20 min were used. Solid matrices such as herbs usually require longer equilibration periods (Fig. 5).

Precision of the analyses

When the analyses are carried out using automatic HSGC equipment, the precision is exceedingly high. Under optimized analytical conditions, C.V. values lower than 1% are usually reached (9). The precision for individual components depends of course on their relative proportions in the material

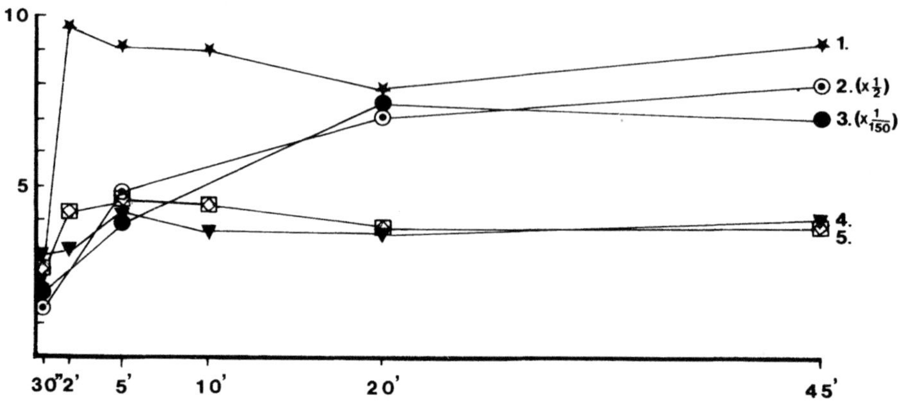

Fig. 5. The effect of equilibration period on the peak areas of some terpenes in solid samples, in HSGC analyses at a bath temperature of 150°C: 1. oxygenated monoterpenes in thyme herb; 2. bisabolol oxide A in chamomile flowerheads; 3. α-bisabolol in chamomile flowerheads; 4. MTHC in thyme herb; 5. sesquiterpenes in thyme herb. Abscissa: equilibration period; ordinate: peak area x 10^4.

and, in HSGC, also on the absolute amounts. Some precision values for a number of components and mixtures are presented in Table I.

Table I. Precision of HSGC. N = number of observations; * = equilibration temperature; □ = 2 samples.

Matrix	Sample	N	C.V. (%)
Liquid	MTHC mixture (60°C)*	5	1.0
Liquid	MTHC mixture (90°C)*	5	0.8
Liquid	MTHC mixture (120°C)*	4	0.3
Liquid	MTHC mixture (150°C)*	5	3.2
Liquid	Thymol	4	2.8
Liquid	Farnesene	6	9.0
Liquid	α-Bisabolol	3	2.8
Liquid	Bisabolol oxide A□	6	1.4
Liquid	Bisabolol oxide A□	6	2.6
Solid	MTHC in larch	6	5.5
Solid	Thymol in thyme herb	4	3.9
Solid	Sesquiterpene hydrocarbons in chamomile flowerheads	6	12.7
Solid	Bisabolol oxide A in chamomile flowerheads□	6	1.7
Solid	Bisabolol oxide A in chamomile flowerheads□	6	1.5

Precision values (C.V.) vary depending on the amounts of the components in HSGC. In some cases the amounts were close to the minimum detectable level of the FID (a few 100 pg). The amounts of some of the components in the MTHC mixture were as high as a few 100 ng. These precision values are as good as, or even better than, the levels obtainable by classical GC using syringe-split/splitless techniques.

Quantitative methods for HSGC of volatile oils

The advantages, disadvantages and applications of quantitative HSGC methods have been discussed in detail in a large number of publications (3, 5, 6, 9-12, 23). The method most widely used for quantitative determination of volatile oils by GC is the so-called proportional quantity method, i.e. the composition is calculated in percentages. This technique is based on the assumption that the detector response for all components is the same, and that all components are vaporizable and elutable. However, this technique does not give the absolute amounts of the individual components because the FID response can vary by as much as 20-30% in comparison to paraffins.

However, the proportional quantity method usually gives sufficient quantitative information for chemotaxonomic studies using HSGC, where large sample series are being screened. The greatest advantage of HSGC in such studies is its speed, since volatile oils can be analyzed directly from fresh or dried samples without prior distillation or extraction. Examples of the use of HSGC in determining chemotypes (*Larix*, *Pinus*, *Matricaria*, *Tanacetum*) are presented in the Figs. 6-7. These fingerprint chromatograms clearly demonstrate the suitability of HSGC for chemotaxonomic work.

Very often more precise quantitative information is needed in volatile oil analysis. As the vapour pressures of the individual components and thus also the equilibrium constants are different, the determination of absolute amounts on the basis of known equilibrium constants or calibration is usually very tedious. In such cases it is better to use normal quantitative GC techniques such as internal and external standardization or the standard addition method. A special quantitative method, multiple headspace extraction (MHE), has also been developed. This method eliminates the matrix effects (9), but it does not compensate for the differences in detector response.

The importance of the matrix effect in quantitative HSGC analyses was

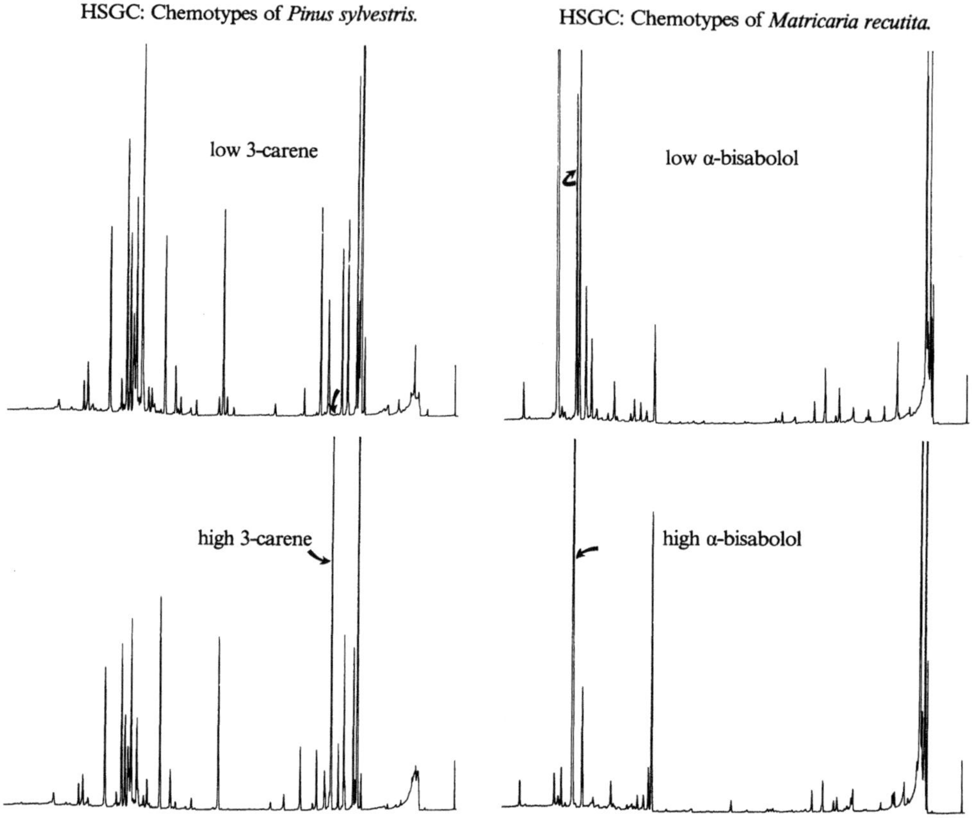

Fig. 6. Chromatograms obtained by HSGC showing two chemotypes of *Pinus sylvestris* (left) and two chemotypes of *Matricaria recutita* (right).

also examined in the present study. When studying the dependence between sample size and the amount of volatile oil of thyme herb in the gas phase, the proportion of the oil was found to decrease from 0.8% to 0.5% (dry weight; range: 0.77-0.48%; s.d.: 0.100; C.V.: 14.9%; N = 7) when the sample size of thyme herb was increased from 30 mg to 70 mg (Fig. 8; left). These determinations were performed using an external standard. The calibration curve was made on the basis of thyme oil (Fig. 8; right). It can be seen from the figures that calibration performed on the basis of thyme oil does not give the right result, especially when the sample size is varied.

In order to eliminate the effect of the matrix on the external standard, we repeated the procedure using 0.04 mg tetradecane as an internal standard.

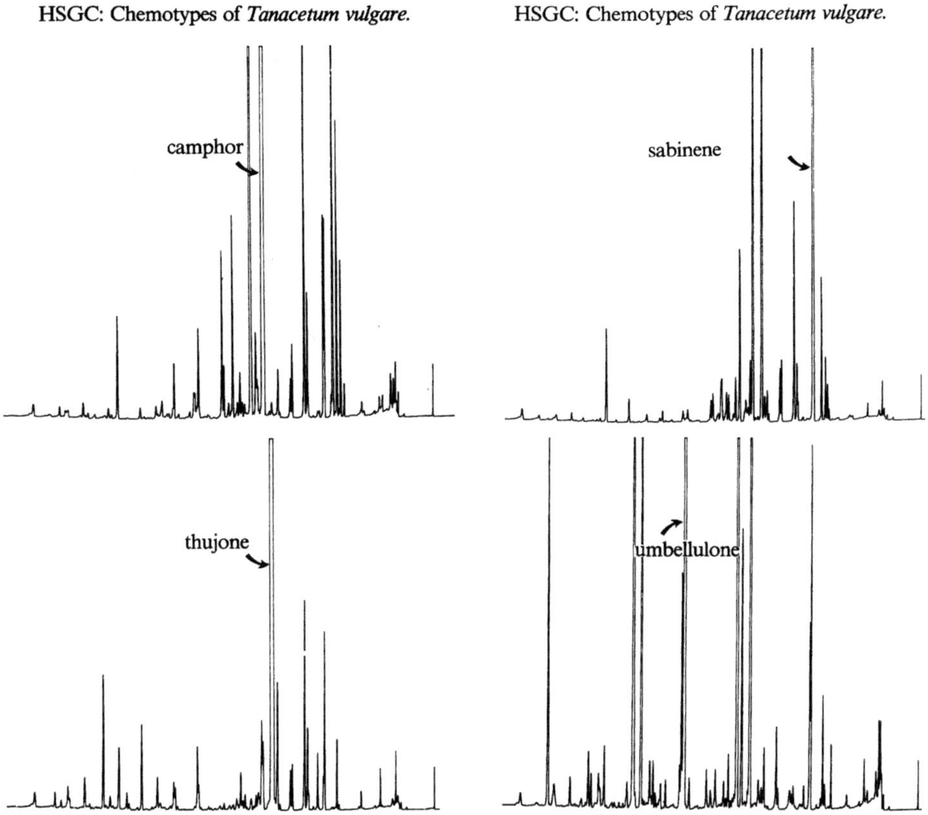

Fig. 7. Chromatograms obtained by HSGC showing four chemotypes of *Tanacetum vulgare*.

A new calibration curve was drawn as a function of the amount of thyme oil added, and the ratio between the total peak area of the volatile oil components and that of the internal standard (Fig. 9). Calibration curve b was formed from curve a by means of the following formula:

$$[A_{tot(added_i)} - A_{tot(0)}] / A_{I.S._i}$$

where i represents the ith addition to the oil.

When the different amounts of thyme herb (Fig. 8; left) were analyzed using this technique, the total amounts of oil obtained from calibration curve b were 0.84% (range: 0.79-0.96%; s.d.: 0.063; C.V.: 0.75%; N = 7). The amount obtained by the officially approved procedure (hydrodistillation and volumetric determination) was 0.8%.

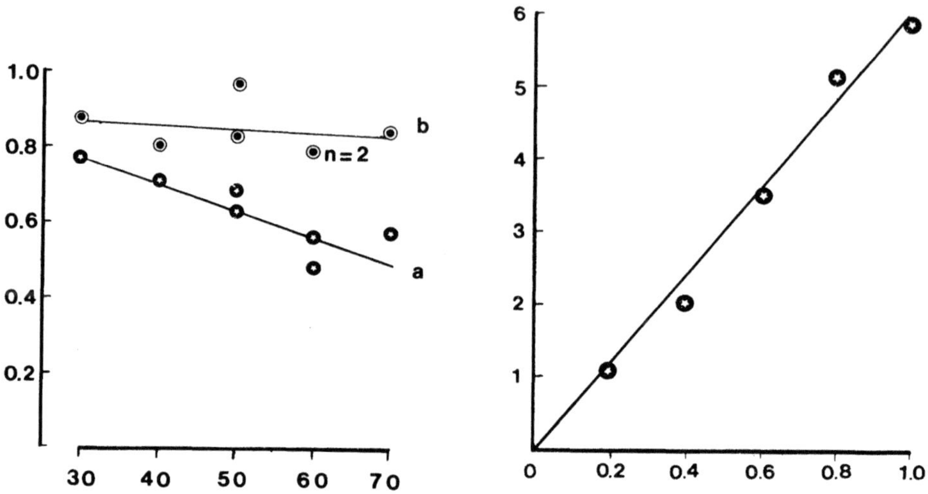

Fig. 8. The volatile oil content of thyme herb obtained by HSGC from various amounts of herb [left]: a. determined using thyme oil as external standard; b. determined using the external standard and the internal standard correction. Abscissa: amount of thyme herb in mg; ordinate: volatile oil content in percentages.
The calibration curve for thyme oil [right]. Abscissa: amount of thyme oil in µl; ordinate: peak area x 10^5.

Fig. 9.

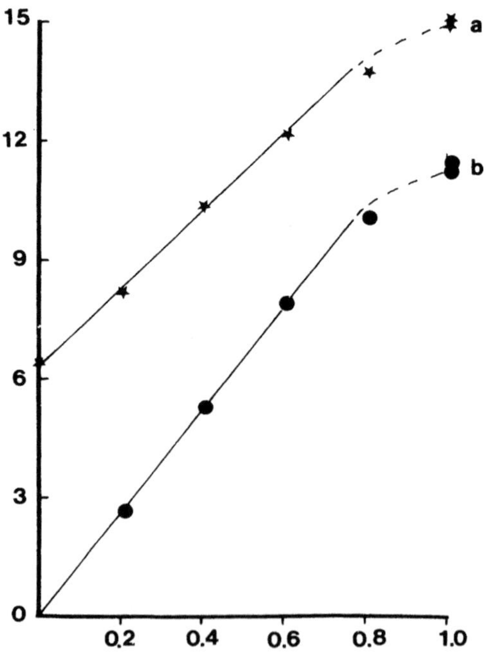

Fig. 9. [left page] Calibration curves for the determination of the amount of volatile oil in thyme herb: a. using thyme oil added to the matrix (thyme herb); b. curve is formed from curve a by calculation using the formula given in the text. Abscissa: added amount of thyme oil in µl; ordinate: peak ratio (A_{tot} / $A_{I.S.}$).

The total amount of oil in thyme herb was also determined using thymol as external standard. In addition, calibration curves were made by adding known amounts of thymol to the herb, and to the herb after having first distilled off the volatile oil (Fig. 10). It can be seen that the slope of curve a (0.50×10^{-6}) differs considerably from the slopes of the curves b and c (0.98×10^{-6} and 1.14×10^{-6} respectively), obtained by the standard addition technique. This is due to the effect of different matrices.

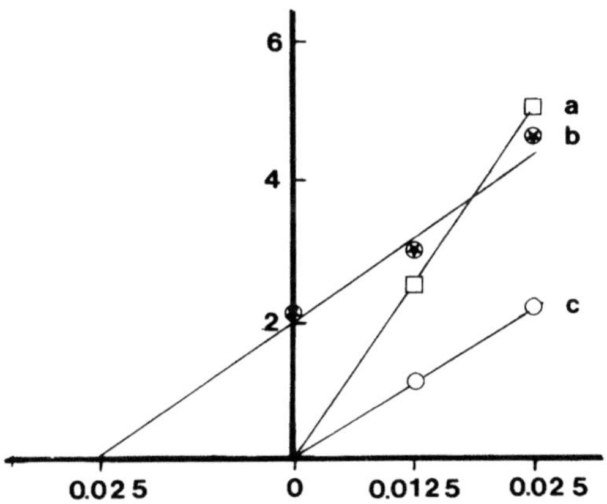

Fig. 10. Calibration curves for the single run HSGC determination of the volatile oil amount in thyme herb: a. thymol as external standard; b. thymol added to the matrix (thyme herb); c. thymol added to the matrix after first having distilled off the volatile oil. Abscissa: added amount of thymol in mg; ordinate: peak area x 10^4.

According to Kolb et al. (9), the matrix effect can be eliminated by using the MHE method. The above experiments and calibration procedures were also carried out using this method (Fig. 11 and Table II). When the slopes of the curves d, e and f (1.70×10^{-7}, 1.64×10^{-7} and 1.47×10^{-7} respectively) are compared, it can be seen that the matrix effects are now approximately the same. The results obtained by different calibration methods are thus acceptable (see also Table II).

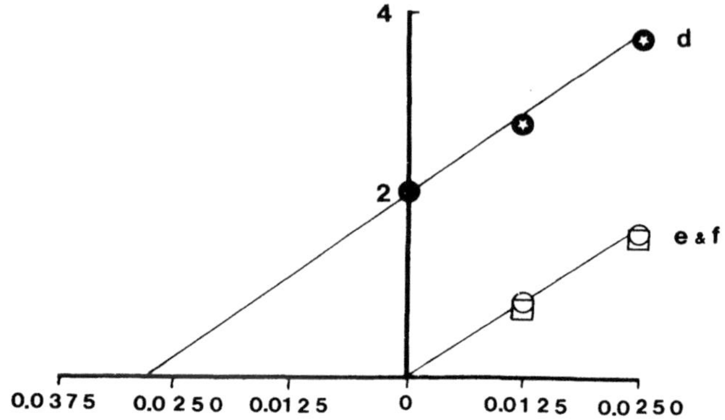

Fig. 11. Calibration curves for the MHE-HSGC determination of the volatile oil amount in thyme herb: d. thymol added to the matrix (thyme herb); e. thymol as external standard; f. thymol added to the matrix after first having distilled off the volatile oil. Abscissa: added amount of thymol in mg; ordinate: peak area x 10^5.

Table II. The total oil content (in percentages) in thyme herb and the amount of thymol (in μg per 30 mg of dried thyme herb) obtained by different calibration methods. a-f, see Figs. 10-11; *pure matrix = thyme herb after first having distilled off the volatile oil.

Calibration method		Oil content	Thymol amount
Single run HSGC analysis			
a	External standard	0.33	10.0
b	Standard addition to matrix (thyme herb)	0.64	19.4
c	Standard addition to pure matrix*	0.74	22.5
MHE-HSGC analysis			
d	Standard addition to matrix (thyme herb)	0.74	31.8
e	External standard	0.74	31.8
f	Standard addition to pure matrix*	0.68	29.2

A higher value is obtained for the amount of thymol using the MHE technique because the vapour pressure of the oxygenated compounds differs considerably from that of the MTHC. So, the relative proportion of p-cymene in single run HSGC was about 28%, whereas that in MHE-HSGC was only 14%. The corresponding values for thymol were 10% and 14%. Owing to the effect of the matrix, addition of the standard to the matrix (thyme herb) gave slightly lower values than addition to the pure matrix in single run HSGC. This also explains the results obtained in the analysis of the flowerheads and of the volatile oil of chamomile. α-Bisabolol was added to both chamomile matrices (flowerheads and oil) to increase the content from 3% to 40%

(Figs. 12-13). It can be seen from the figures that this resulted in a decrease of the absolute and relative amounts of the different components in the material. This was presumably also the case when thymol was added to the thyme herb. Addition of the standard resulted in an apparent decrease in the relative proportion of thymol in the sample. The MHE method partly compensates for this source of error (Table II).

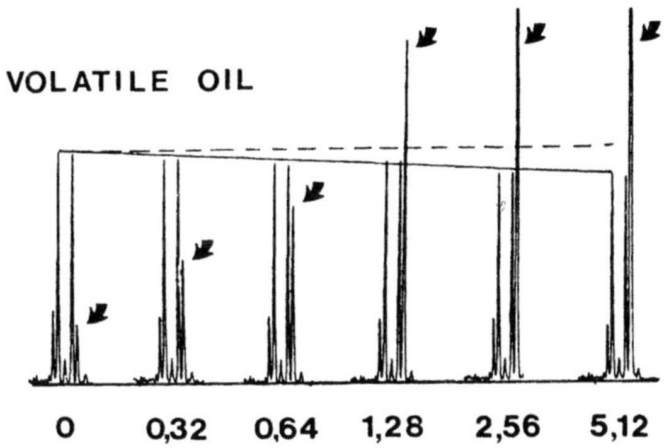

Fig. 12. The effect of the matrix on the amounts of the components in chamomile oil after addition of various amounts of α-bisabolol (arrow). Abscissa: added amount of α-bisabolol x 10^{-2} mg.

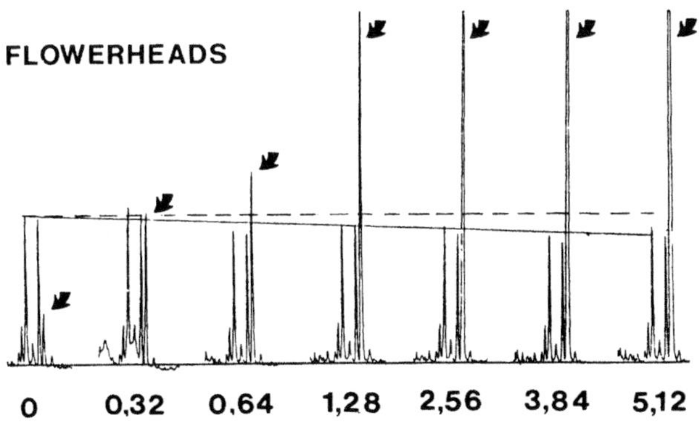

Fig. 13. The effect of the matrix on the amounts of the components in chamomile flowerheads after addition of various amounts of α-bisabolol (arrow). Abscissa: added amount of α-bisabolol x 10^{-2} mg.

The effect of the matrix on quantitative HSGC analyses was also studied using some commercially available chamomile preparations. An artificial matrix, ethanol-water 30:70 (w/w), was spiked with known amounts of α-bisabolol and a new calibration curve was made (Fig. 14). A calibration

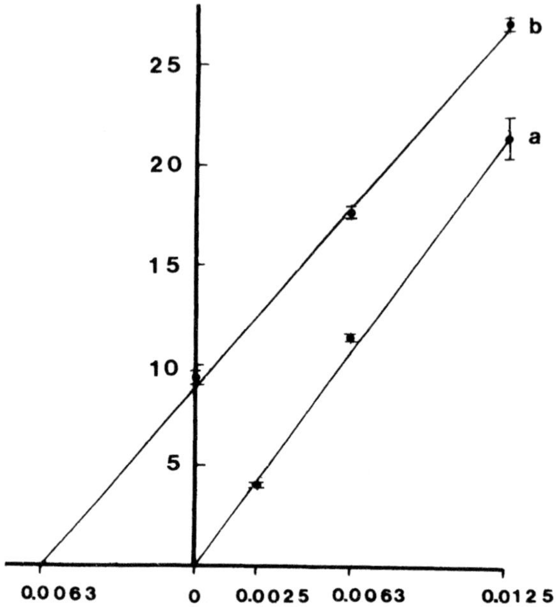

Fig. 14. Determination of the amount of α-bisabolol in a commercially available chamomile preparation: a. α-bisabolol added to an artificial matrix, ethanol-water, N = 3; b. α-bisabolol added to a chamomile preparation, N = 6. Abscissa: added amount of α-bisabolol in mg; ordinate: peak area x 10^3.

curve was also made by adding known amounts of α-bisabolol to one of the chamomile preparations (Fig. 14). The external standard gave an α-bisabolol content for the preparation of 53.3 mg/100 g (range: 51.7-56.7 mg/100 g; s.d.: 5.700; C.V.: 10.7%; N = 6). The standard addition technique gave 62.5 mg/100 g (range: 58.8-67.4 mg/100 g; s.d.: 3.325; C.V.: 5.3%; N = 6). Both methods take the detector response into account.

The standard addition method takes the effect of the matrix on the equilibrium constant of α-bisabolol into account. On the other hand, when the external standard addition method is used, the artificial matrix is not fully comparable with that of the chamomile preparation (slope of curve a: 5.81 x 10^{-7}; that of curve b: 7.24 x 10^{-7}). The external standardization and MHE methods gave a value for the α-bisabolol content of 63.0 mg/100 g (mean of two determinations).

HSGC analysis of MTHC in both solid and liquid matrices is exceedingly straightforward, because they have similar vapour pressures and equilibrium constants, as can be seen from the results obtained by MHE (Fig. 15).

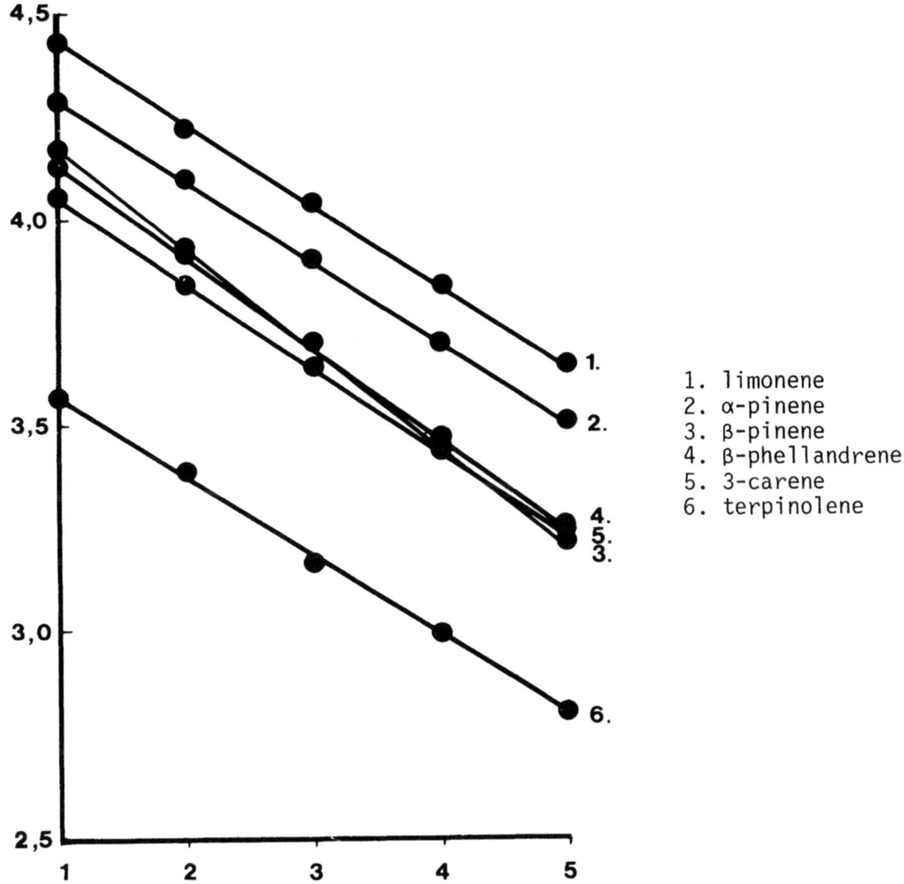

Fig. 15. MHE-HSGC analysis of some monoterpene hydrocarbons. Abscissa: number of extractions; ordinate: log peak area.

These compounds rapidly reach a state of equilibrium even at relatively low temperatures. The precision of the analyses is lower when high equilibration temperatures are used, due to the fact that the pressure in the vials increases rapidly and leaks may ensue (see also Table III).

The relative proportions of the different MTHC did not differ much at different bath temperatures (Table III). They also closely resembled the results obtained with the classical GC split technique. The greatest differences occurred in MHE-HSGC analyses with β-pinene and terpinolene, which have different equilibrium constants (see Fig. 15). The equilibrium of β-pinene in the gas phase is larger, that of terpinolene smaller.

Table III. The percentage composition of the MTHC mixture obtained at different bath temperatures and using single run HSGC and MHE-HSGC. The results of classical GC are given as reference.

Compound	Single run HSGC Bath temperature 60°C	90°C	120°C	150°C	MHE-HSGC 150°C	GC split
α-Pinene	20.9	21.2	21.4	21.2	20.3	21.1
β-Pinene	19.7	19.9	20.0	19.4	17.9	20.3
3-Carene	12.0	11.9	12.0	12.1	12.0	12.2
Limonene	28.2	28.4	28.2	28.1	28.9	27.6
β-Phellandrene	15.5	14.9	14.6	15.6	16.0	15.6
Terpinolene	3.8	3.8	3.8	3.8	5.0	3.5
C.V. in %	1.0	0.8	0.3	3.2	11.5	
Number of runs	5	5	3	5	5	

The relative proportions of some MTHC determined in thyme oil and in thyme herb by means of GC (split technique), and by single run HSGC as well as by MHE-HSGC respectively are presented in Table IV.

Table IV. The percentage composition of the MTHC mixture in thyme herb determined by single run HSGC and MHE-HSGC, and in thyme oil by GC (split technique).

Compound	Single run HSGC	MHE-HSGC	GC split
α-Pinene	3.3	2.8	2.0
Camphene	5.5	4.8	3.5
Sabinene	2.0	2.5	1.8
Limonene	9.2	10.0	7.0
γ-Terpinene	2.7	2.8	3.0
p-Cymene	77.2	77.1	82.7

The results of HSGC do not vary much. On the other hand, they do differ from those obtained by GC of the volatile oil sample. The large difference in the case of p-cymene is due to the relative decrease of the other components during the hydrodistillation used for isolation of the oil. It may also be due to the discrimination effect of the split method on the other components.

REFERENCES

(1) Núñez AJ, González LF, Janak J. 1984, J. Chromatogr. 300:127.
(2) McAuliffe C. 1971, Chem. Techn. 1:46.
(3) Kolb B, Auer M, Pospisil P. 1981, Applied Chromatography No. 35E, Ueberlingen, Bodenseewerk Perkin-Elmer & Co. GmbH.
(4) Kolb B, Pospisil P. 1977, Chromatographia 10:705.

(5) Kolb B. 1982, Chromatographia 15:587.
(6) Kolb B. 1976, J. Chromatogr. 122:553.
(7) Kolb B, Auer M, Pospisil P. 1981, J. Chromatogr. 204:371.
(8) Ioffe BV, Vitenberg AG. 1978, Chromatographia 11:282.
(9) Kolb B, Auer M, Pospisil P. 1983, J. Chromatogr. 279:341.
(10) Drozd J, Novak J. 1979, J. Chromatogr. 165:141.
(11) Drozd J, Novak J. 1984, J. Chromatogr. 285:478.
(12) Hochenberg HH, Schmidt AP. 1977, Gas Chromatographic Headspace Analysis, New York, Heyden & Son, p. 10.
(13) Roeraade J, Blomberg S. 1983, Chromatographia 17:387.
(14) Miyaura S, Mikawa R, Isono H. 1983, Eisei Kagaku 29:83.
(15) Lee KY, Nurok D, Zlatkis A. 1978, J. Chromatogr. 158:377.
(16) Hiltunen R, Löyttyniemi R, Räisänen S, Tigerstedt PMA. 1982, Silva Fennica 16:231.
(17) Chialva F, Doglia G, Gabri G, Ulian F. 1983, J. Chromatogr. 279:333.
(18) Gabri G, Chialva F. 1981, J. High Resol. Chromatogr. & Chromatogr. Commun. 4:216.
(19) Chialva F, Gabri G, Liddle PAP, Ulian F. 1982, J. High Resol. Chromatogr. & Chromatogr. Commun. 5:182.
(20) Hiltunen R, Laakso I, Scheffer JJC. 1983, Acta Pharm. Fenn. 98:209.
(21) Riba ML, Randrianalimane, Torres L, Mathieu J. 1983, Chromatographia 17:497.
(22) Hiltunen R, Räisänen S, Von Schantz M. 1980, Planta Med. Suppl.:112.
(23) Raccio JM, Widomski JR. 1981, Chromatography Newsletter 9:42, Perkin-Elmer.

CONCENTRATION OF ODOROUS HEADSPACE VOLATILES

R. TER HEIDE

ABSTRACT

Highly volatile odorous compounds were studied by examination of the vapour phase above samples by so-called headspace analysis. The static method was applied to study equilibria as they occur in alcoholic beverages. To obtain detectable amounts of minor components the headspace was enriched. Transfer of the vapours with an inert gas onto the porous polymers Chromosorb 102 and Tenax GC was found to be suitable for this purpose. Methods to concentrate different amounts of headspace vapour are described. The volatiles were thermally desorbed and condensed in a glass capillary tube to enable odour evaluation, GC-MS analysis, IR and NMR spectroscopy. Interfering amounts of water could be avoided by using a tandem-wise arrangement of traps taking advantage of the low breakthrough volume of water. Ethanol and monoterpene hydrocarbons could be retained on specific stationary phases.

Results on headspace analysis of alcoholic beverages, clary sage oil and spike lavender oil are reported.

INTRODUCTION

People concerned with the investigation of complex natural odorous products have learned that knowledge of chemical compounds responsible for the odour character cannot be achieved by a single treatment of the sample. To study the volatile constituents, examination of the vapour phase above a sample by so-called headspace techniques was found to be very attractive. Laborious isolation techniques applied to the sample itself are avoided and clean samples for analysis can be prepared. In this paper headspace techniques as they were developed in our laboratory for identification of odorous components of natural products are discussed. The methods were applied in the analysis of alcoholic beverages and essential oils.

DIRECT HEADSPACE ANALYSIS

The concentration of a compound in the vapour phase is roughly a function of its vapour pressure at a given temperature, the concentration and the degree of interaction with other substances present in the sample.

A. Baerheim Svendsen and J.J.C. Scheffer (eds), Essential oils and aromatic plants, ISBN-13: 978-94-009-5137-2

For investigation of the vapour phase two methods can be distinguished, the static and the dynamic headspace method.

The static or direct method is the simplest one. Advantages can be summarized as follows:

- a relative small sample is required;
- sample preparation is very simple;
- artifact formation is kept to a minimum;
- quantitative analysis of major components in the vapour phase is relatively simple and accurate;
- equilibria can be studied such as those between aldehydes and acetals and between acids and their ethyl esters in ethanolic solutions as occurring in alcoholic beverages;
- vapour phases of different products can be 'fingerprinted' for quality control purposes;
- automation is possible.

As a part of our analytical work on alcoholic beverages we applied the static headspace method to study equilibria between aldehydes and diethylacetals in ethanolic solutions. The reactions involved are given in Fig. 1.

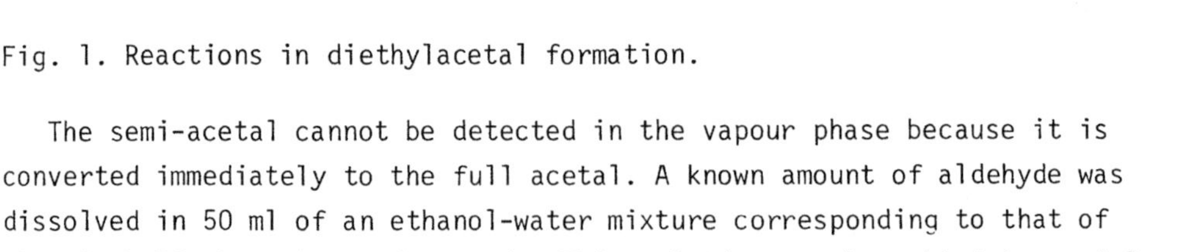

Fig. 1. Reactions in diethylacetal formation.

The semi-acetal cannot be detected in the vapour phase because it is converted immediately to the full acetal. A known amount of aldehyde was dissolved in 50 ml of an ethanol-water mixture corresponding to that of the alcoholic beverage under study. This solution was brought into a vial of 100 ml covered with an aluminium cap provided with a pierceable teflon lined septum. The vial was placed in a thermostated bath at 25°C and, at equilibrium, 3 ml of the vapour phase was withdrawn with a gas-tight syringe previously cleaned and conditioned at the sample temperature. The vapour sample was analyzed by gas chromatography.

A linear relation should exist between the concentration of the aldehyde and its peak area. At pH 7 no acetal is formed. Lowering of the pH to a value corresponding to that of the beverage under investigation and determination of the aldehyde concentration at different time intervals gives an impression of the speed of acetal formation. Aldehyde-acetal equilibration can best be studied at pH 1, at which the balance is reached within a few minutes.

Our work on alcoholic beverages revealed that in whisky and cognac of 40% (v/v) alcohol content, 15% of the lower aldehydes is converted to the corresponding diethylacetals. In rum, containing 75% (v/v) of ethanol this conversion is 25%, expressed as moles/l.

The concentration of formic acid in alcoholic beverages can be determined indirectly by studying the equilibrium between formic acid and ethyl formate in ethanolic solution (Fig. 2).

$$H-COOH + C_2H_5OH \rightleftharpoons H-COOC_2H_5 + H_2O$$

Fig. 2. Equilibrium formic acid - ethyl formate in ethanol.

Direct vapour analysis has also some major disadvantages:
- identification by spectroscopic methods is limited to the main component in the vapour phase;
- detailed odour evaluation of individual components is not possible.

HEADSPACE ENRICHMENT

The vapour phase of a natural product may contain many odorous compounds in relatively low concentrations. For their identification at least GC-MS is required. If the mass spectra do not give decisive information on the structure of a compound, enough sample must be availabe te permit sniffing of the unknowns at the outlet of a capillary GC column to judge the olfactive importance, an to permit trapping of the compounds of interest for IR and NMR analysis. Of the various methods used for headspace enrichment, sweeping of the volatiles with an inert gas onto porous polymer traps was found to be most suitable.

Advantages of headspace enrichment are:
- odour evaluation of the concentrate and the individual components is possible, as well as

- spectroscopic analysis of individual compounds;
- dominant vapour compounds such as water and ethanol can be removed by selective adsorption.

A disadvantage of the method is that equilibrium can be disturbed when large amounts of vapour are withdrawn. In qualitative analysis this is not a serious drawback.

The sorbent should meet the following requirements:

- high adsorptive capacity for flavour and fragrance volatiles at relatively high gas flows;
- low affinity for dominant compounds such as water and ethanol;
- easy release of adsorbed compounds;
- no chemical or catalytic activity;
- no release of blank impurities;
- thermally stable;
- unshrinkable.

Chromosorb 102 and Tenax GC reasonably fulfil these requirements.Chromosorb 102 is a styrene-divinylbenzene co-polymer with a surface area of 300-400 m^2/g. The temperature limit is 250^oC. Before each experiment it was conditioned overnight at 180^oC in a Nitrogen stream of 300 ml/min. Recovery of high boiling compounds with more than 10 carbon atoms may give problems.

Tenax GC is poly-(2,6-diphenyl-p-phenyleneoxide) with a surface area of 20-30 m^2/g. It has a temperature limit of 375^oC and was freshly conditioned overnight at $200\text{-}250^oC$ in a Nitrogen stream of 300 ml/min. Particle size was 60-80 mesh. Tenax GC is preferred when less volatile compounds have to be concentrated because higher desorption temperatures can be used. Table I shows some breakthrough volumes in ml/g adsorbent at 25^oC.

Table I. Breakthrough volumes of some volatiles on Chromosorb 102 and Tenax GC in ml/g at 25^oC.

	Chromosorb 102	Tenax GC
Water	110	60
Methanol	590	60
Ethanol	2400	300
Isopentanol		2300
Ethanal	540	
Ethyl acetate	4500	6000
Ethyl butyrate		250000
Dimethyl sulfide	5000	

The breakthrough volumes were measured by installing the adsorption tubes in a preparative gas chromatograph and the first detection of the peak front was defined as breakthrough volume. Both adsorbents have low affinity for water. Desorption volumes at higher temperatures used in the desorption steps were measured at the point where the last trace of compound could be detected.

Nitrogen stripping gas was obtained from evaporating liquid Nitrogen. Helium used as carrier gas in desorption was taken from a cylinder and this gas was dried by passage through an U-shaped tube packed with 2 g of freshly conditioned (100°C) gelatine powder and held at -196°C.

The apparatus used in headspace concentration is relatively simple of construction. The dimensions of the sample container are not critical. A high surface area was created by bringing the sample in a cylindrical container (60 cm x 10 cm) of Quick-fit glass. It can be used for stripping of the headspace of beverages, fruits, etc. Carrier gas was introduced on one side and the sample vapours were conducted into an adsorbent tube connected at the other side of the container. A simple round-bottomed flask provided with a gas inlet and a connector to attach an adsorbent tube, served as container if percolation of the sample with transport gas was required. The sample containers could be heated with Calorex heating tube.

The adsorbent tubes were of glass. Larger amounts of adsorbent were brought in a spiralized tube (3 m x 8 mm i.d.), which could contain 45 g of Chromosorb 102. Smaller amounts of adsorbent were filled in straight glass tubes (15 cm). Their diameter was adapted to the amount of adsorbent used. The narrowed ends of all glass tubes were threaded externally. Connection of tubes is shown schematically in Fig. 3.

The tubes were separated by a thin teflon spacer to avoid cracking, and a polyimide nut ensured a leak-free seal. Thermal desorption of the spiralized glass tube was performed in an adapted drying oven. The straight glass tubes were placed in an aluminium block, which could be heated with cartridge heaters.

The concentration of different amounts of headspace vapours will now be described. In Fig. 4 the enrichment of 10 l of vapour is depicted. First 10 l of the stripping gas, Nitrogen, is swept over the sample with a flow of 100 ml/min and led onto trap A containing 45 g of Chromosorb 102, held at room temperature.

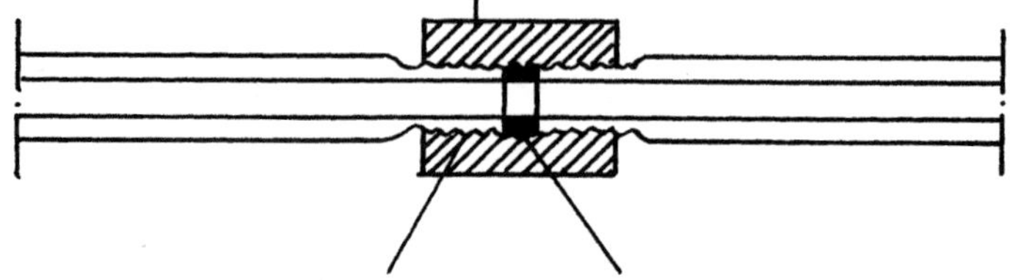

Fig. 3. Connection of adsorbent tubes.

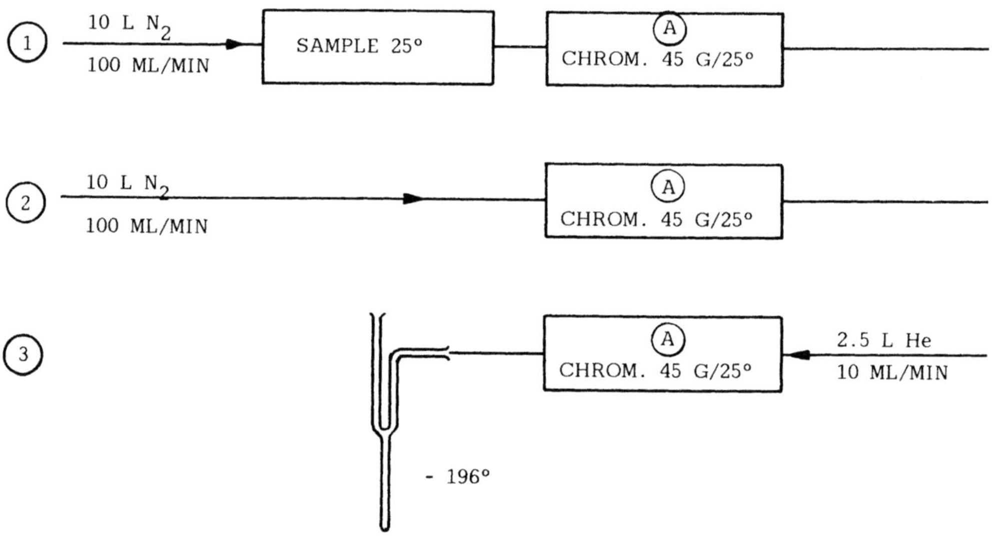

Fig. 4. Concentration of 10 l of headspace vapour. Breakthrough volumes: acetaldehyde, 24.3 l/45 g of Chromosorb 102 (25°C); water 4.9 l/ 45 g of Chromosorb 102 (25°C). Desorption volume: acetaldehyde 1.8 l/45 g of Chromosorb 102 (150°C).

Suppose we want to retain acetaldehyde on the adsorbent. According to its breakthrough volume, passing of 24.3 l Nitrogen gas through the trap will cause no breakthrough of this aldehyde. It follows that the trap can be washed in the second step with 14.3 l of pure Nitrogen to remove adsorbed water. We preferred to wash with 10 l of Nitrogen as a safe volume.

Finally the volatiles were desorbed at 150°C by leading 2.5 l of pure and dry Helium in reverse direction through the Chromosorb trap A at a flow of 10 ml/min. This is amply sufficient to elute acetaldehyde and the other volatiles completely. The desorbed volatiles are condensed in a Y-shaped glass capillary tube of 1 mm i.d. at -196^{0}C.

The same arrangement can be used if we want to collect 90 l of headspace vapour (Fig. 5).

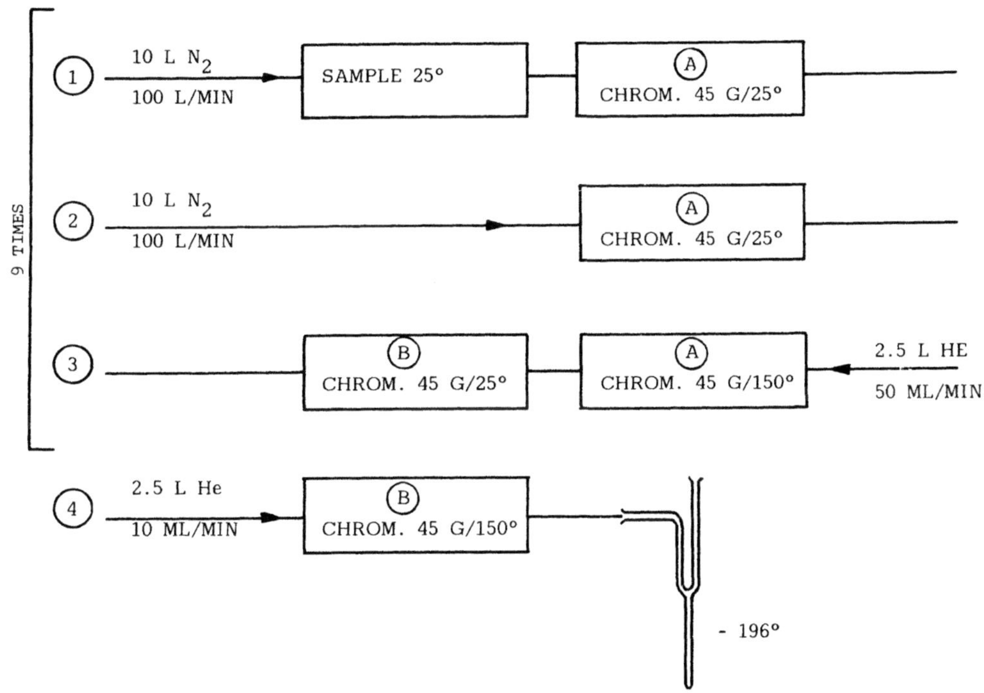

Fig. 5. Concentration of 90 l of headspace vapour. Breakthrough volumes and desorption volume: see Fig. 4.

Now the desorbed volatiles were not condensed in the capillary tube but transferred onto another trap B, containing also 45 g of Chromosorb 102. This trap, held at room temperature, can take up 24.3 l of desorption gas before breakthrough of acetaldehyde, so the desorption step 3 can be repeated nine times with 2.5 l of Helium. In step 4, the trap B containing 90 l of enriched vapour was desorbed in reverse direction at 150^{0}C with 2.5 l of Helium and the volatiles were condensed in a glass capillary, cooled in liquid Nitrogen.

It was found in practice that not always a water-free condensate was obtained. To eliminate residual water, the method shown in Fig. 6 was applied.

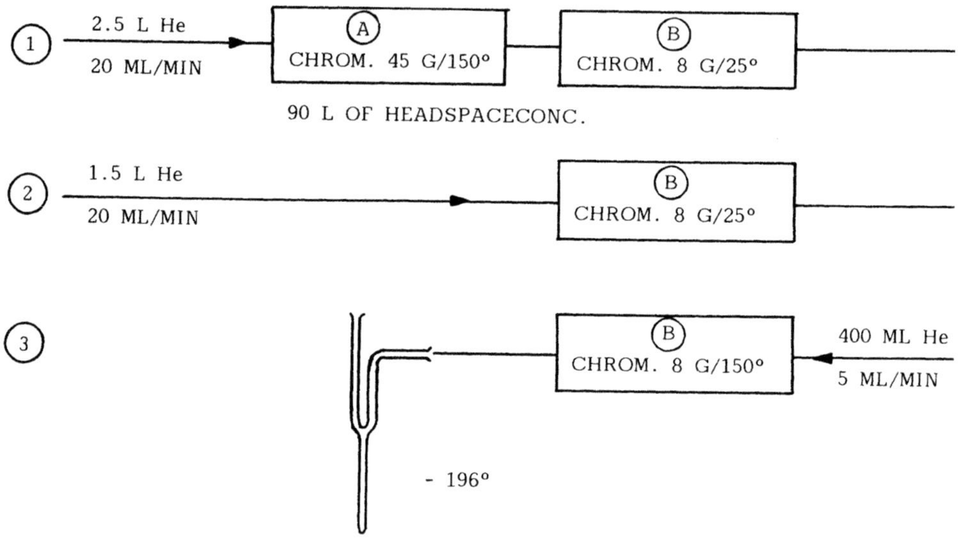

Fig. 6. Water-free collection of 90 l of headspace volatiles. Breakthrough volume of acetaldehyde (Chromosorb 102): 4.3 l/8 g (25° C). Desorption volumes: 1.8 l/45 g (150°C); 32 ml/8 g (150°C).

Trap A containing 90 l of enriched headspace volatiles, obtained as described before, was coupled with a small trap containing 8 g of Chromosorb 102. The trap A was desorbed at 150°C and the volatiles were transferred to trap B, at room temperature. The breakthrough volume of acetaldehyde is 4.3 l at 25°C. To remove residual water this tube can be washed with 1.5 l of pure, dry Helium in the same direction. Desorption at 150°C was performed in reversed direction with 400 ml of Helium as a safe volume. The applied low flow of 5 ml/min guaranteed a high collection efficiency in the glass capillary.

For alcoholic beverages we found (1) that ethanol can occur as an interfering component in the condensate, obtained by the method shown in Fig. 6. Ethanol was not removed completely because of tailing effects in the adsorption step. Introduction of a spiralized glass tube (2 m x 6 mm i.d.) filled with 5 g of diglycerol coated on 15 g of Embacel (60-80 mesh), permitted removal of residual amounts of ethanol (Fig. 7). Its removal was most effective if the enriched headspace concentrate was introduced on the top of the diglycerol column as a small plug.

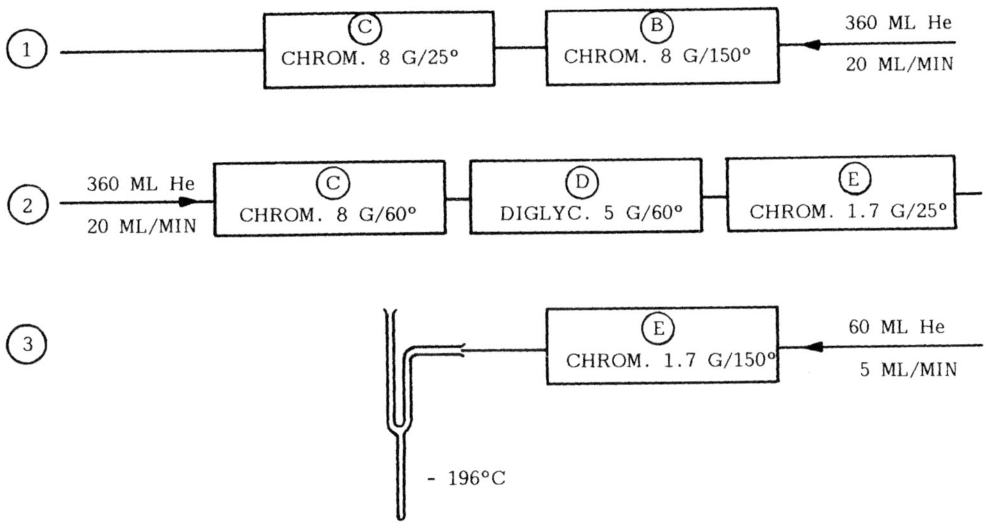

Fig. 7. Ethanol-free collection of headspace volatiles. Breakthrough volumes: ethanol, 380 ml/5 g of diglycerol (60°C); acetaldehyde, 920 ml/1.7 g of Chromosorb 102 (25°C). Desorption volumes: acetaldehyde, 320 ml/8 g of Chromosorb 102 (150°C); acetaldehyde, 60 ml/1.7 g of Chromosorb 102 (150°C).

Column B containing a concentrate of 90 l of headspace volatiles was desorbed at 150°C and connected to an identical trap C, held at room temperature. Adsorption takes place on the top of this column C. In a second step this trap was backflushed at 60°C with 360 ml of Helium and the volatiles were introduced as a small plug into the diglycerol column D at 60°C. Ethanol breaks through this column with 380 ml of Helium, so it was retained completely. The volatiles leaving the diglycerol column were collected on a small trap E, at room temperature, containing 1.7 g of Chromosorb 102, from which they were backflushed with 60 ml of Helium at 150°C into the glass trap at -196°C. Esters and ketones with more than eight carbon atoms are also retained on the diglycerol column.

If collection of still larger amounts of vapour is required to detect less volatile compounds, then Tenax GC is preferably used as porous polymer. It has higher thermal stability than Chromosorb 102, enabling elution of less volatile compounds at higher desorption temperatures. Fig. 8 shows how we concentrated 1000 l of headspace vapour.

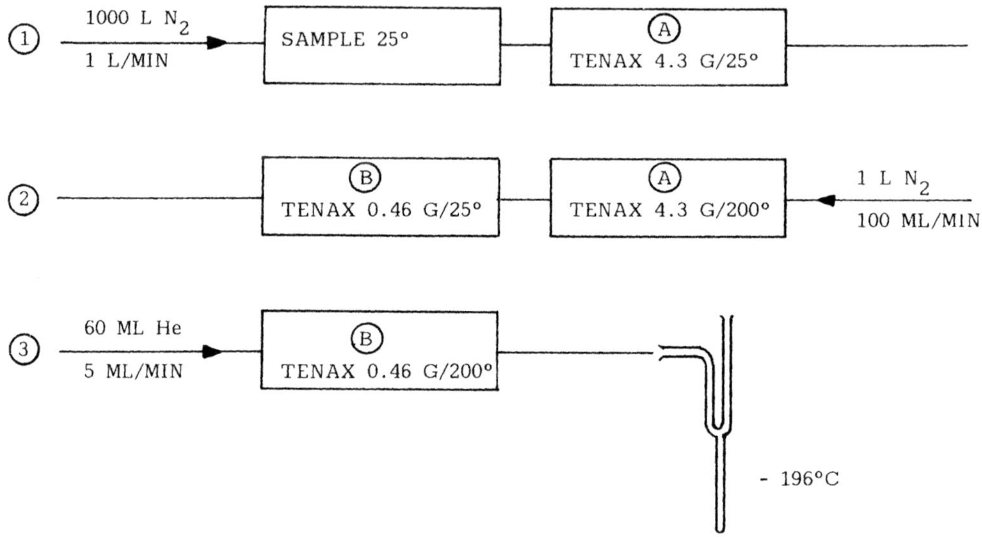

Fig. 8. Concentration of 1000 l of headspace vapour. Breakthrough volumes: water, 258 ml/4.3 g of Tenax GC (25°C); ethyl butyrate, 1075 l/ 4.3 g of Tenax GC (25°C); ethyl butyrate, 115 l/0.46 g of Tenax GC (25°C). Desorption volume: ethyl butyrate, 100 ml/4.3 g of Tenax GC (200°C).

Compounds such as water and ethanol do not interfere, because they are eluted completely from the trap. Because equilibrium conditions cannot be maintained under the conditions applied, this method is only suitable for qualitative work. The transport gas was swept over the sample with a flow of 1 l/min onto trap A, containing 4.3 g of Tenax GC. The sample was refreshed at regular intervals.

Ethyl butyrate, having a breakthrough volume of 1075 l on trap A at room temperature, remains just on the adsorbent. Trap A was desorbed at 200°C in reverse direction with 1 l of Nitrogen and the volatiles were transferred onto a smaller trap with 0.46 g of Tenax GC. This trap B was introduced to improve the yield of condensation in the capillary. In the third step only 60 ml of Helium was needed to desorb the enriched volatiles from trap B at 200°C.

A review of the literature revealed that, surprisingly, the highly volatile components of essential oils have received relatively little attention. This may be due to the fact that they occur in low concentrations as compared with the other oil constituents. They easily escape attention during chromatography of the oils, a technique frequently applied for separation of the various groups of essential oil constituents having different polarity.

In 1971 we published about the highly volatile components of geranium oil (2). They were isolated by distillation with carbon dioxide and by spinning band distillation. Such compounds can also be studied by headspace analysis. Interference of monoterpene hydrocarbons in enriched headspace samples was overcome by selective adsorption. In spite of the fact that theoretically α-pinene should be quantitatively adsorbed on porous polymers, due to the large excess, part of the monoterpenes was carried along with the stripping gas. Fig. 9 shows the method developed for concentration of the headspace of essential oils.

First 20 l of Nitrogen was passed over or through the oil at a flow of 100 ml/min. The volatiles were collected on 45 g of Chromosorb 102 at room temperature. Acetaldehyde was retained. Secondly this column A was backflushed at 150°C with 4 l of Nitrogen onto a smaller trap B, containing 8 g of Chromosorb 102. This column was coupled with a spiralized column (2 m x 8 mm i.d.) packed with 2.5 g of purified Apiezon-L coated on 25 g of silanized Embacel (60-80 mesh). Volatiles desorbed from trap B at 150°C were transferred onto this column C with 360 ml of Helium at a flow of 20 ml/min. The column was heated to 120°C. At this temperature the breakthrough volume of α-pinene is 400 ml, which means that all monoterpene hydrocarbons were retained. The non-retained volatiles were collected on 1.5 g Chromosorb 102 (trap D) at room temperature From this trap they were backflushed at 150°C in reverse direction and condensed at -196°C. Instead of Chromosorb 102 also Tenax GC can be used, provided that breakthrough volumes are taken into account.

Quantitative analysis was already discussed in connection with direct headspace techniques. Calibration curves of components of which the vapour concentration is too low to permit detection, can be prepared after enrichment of the components.

To demonstrate the usefulness of this method, we compared the relative peak areas of a number of esters measured by direct vapour analysis as well as by concentration of 50 l of the vapour phase.

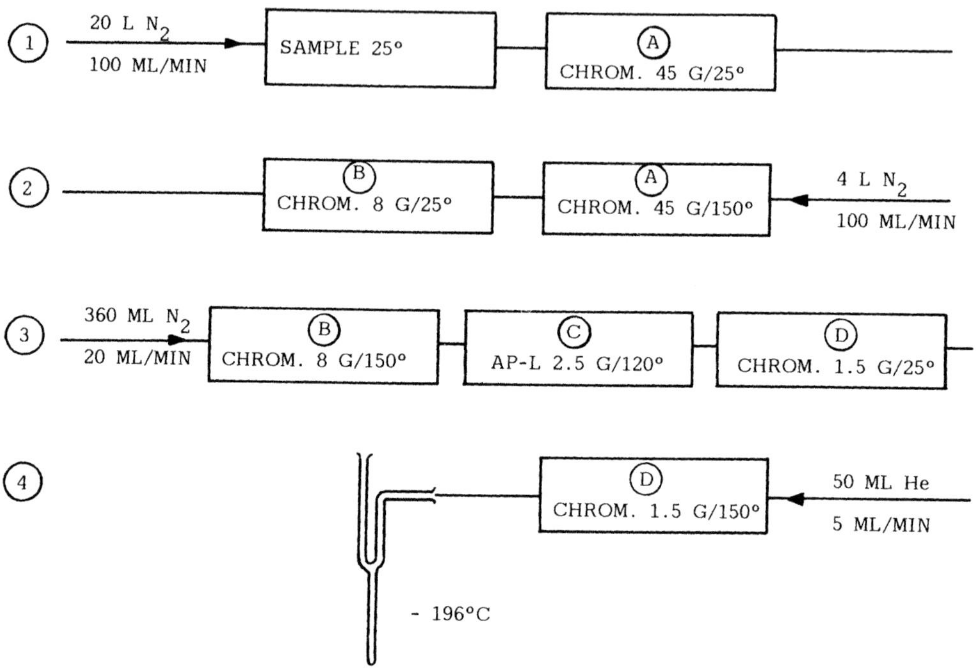

Fig. 9. Hydrocarbon-free collection of headspace volatiles of essential oils. Breakthrough volumes (in addition to those mentioned in Figs. 4 and 6): α-pinene, 400 ml/2.5 g of Apiezon-L (120°C); acetaldehyde, 800 ml/1.5 g of Chromosorb 102 (25°C). Desorption volume: acetaldehyde, 50 ml/1.5 g of Chromosorb 102 (150°C)

Table II lists the peak areas relative to that of ethyl butyrate (internal standard) determined at 25°C. A sucrose solution as applied in lemonade production was chosen as matrix. The concentration of each ester was 2 ppm. In direct vapour analysis 0.5 ml of vapour above 25 ml of the matrix was analyzed by GC and the peak areas were determined using ethyl butyrate as internal standard.

The vapour phase was also enriched by leading 50 l of Nitrogen at a flow of 200 ml/min over 2 l of the same ester mixture in a cylindrical sample container. The vapour was collected in 0.9 g of Tenax GC and this trap was washed with 4 l of dry Helium to remove water. In reverse direction the esters were desorbed at 200°C with 60 ml of Helium at a flow of 5 ml/min. They were condensed in a Y-shaped capillary held in liquid Nitrogen. The condensate was analyzed by GC and the peak areas were compared with that of ethyl butyrate.

Table II. Comparison of headspace concentrations of some esters above a sucrose solution as determined by direct sampling and by enrichment on Tenax GC. Matrix: sucrose solution 67°Brix, 125 ml; citric acid, 50% solution, 2.5 ml; water, 872.5 ml; ester concentration, 2 ppm.

	Direct sampling, 25°C	Enrichment of 50 l vapour on Tenax GC, 25°C
Heptyl acetate	5.5	5.1
Benzyl acetate	0.2	0.15
Benzyl butyrate	0.6	0.6
Ethyl heptylate	3	4
Methyl decylate	13	12
Ethyl benzoate	0.5	0.7
Methyl salicylate	0.3	0.4
Ethyl butyrate (internal standard)	1	1

It could be concluded that the equilibrium was not disturbed in the enrichment procedure and that reasonable agreement existed between both headspace methods. The table shows also the dependence of the headspace concentration in relation to the chemical structure. Methyl salicylate and methyl decylate, having the same boiling point, have markedly different vapour concentration in this matrix.

APPLICATIONS

The techniques described were found to be very powerful in identifying volatiles in the headspace of alcoholic beverages (3). Table III gives a survey of the compounds detected in the headspaces of rum, cognac and whisky.

Table III. Headspace compounds of rum (R), cognac (C) and whisky (W).

		R	C	W
ACETALS				
	Diethoxymethane	+	+	
	Diethoxyethane	+	+	+
	1-Ethoxy-1-propoxyethane		+	
	1-Ethoxy-1-(2-methylpropoxy)ethane		+	
	1-Ethoxy-1-(2-methylbutoxy)ethane		+	
	1-Ethoxy-1-(3-methylbutoxy)ethane		+	
	1-Ethoxy-1-hexoxyethane		+	
	1,1-Diethoxypropane		+	
	1,1-Diethoxy-2-methylpropane	+	+	+

Table III (continued)

		R	C	W
	1,1-Diethoxy-3-methylbutane	+	+	+
	1,1-Diethoxyhexane			+
	1,1-Diethoxynonane			+
ALDEHYDES				
	Ethanal	+	+	+
	Propanal	+	+	
	1-Butanal	+		
	Isobutanal	+	+	+
	1-Pentanal	+		
	2-Methylbutanal	+	+	
	3-Methylbutanal	+	+	+
	1-Hexanal	+		+
	1-Heptanal	+		+
	1-Nonanal	+		+
	2-Methyl-2-propen-1-al		+	
	Benzaldehyde	+	+	
	Furfural	+	+	+
ALCOHOLS				
	Methanol	+		
	Ethanol	+	+	+
	Propanol	+	+	+
	Butanol	+	+	
	2-Butanol	+	+	
	Isobutanol	+	+	+
	Pentanol		+	+
	3-Pentanol		+	
	2-Methylbutanol	+	+	+
	3-Methylbutanol	+	+	+
	Hexanol		+	
	2-Phenylethanol		+	+
ESTERS				
	Methyl formate		+	
	Ethyl formate	+	+	+
	Isobutyl formate		+	
	Isopentyl formate		+	
	Methyl acetate		+	
	Ethyl acetate	+	+	+
	Propyl acetate	+	+	
	Isobutyl acetate		+	
	Isopentyl acetate		+	
	Hexyl acetate		+	
	Ethyl propionate	+	+	+
	Ethyl butyrate	+	+	+
	Butyl butyrate		+	
	Isopentyl butyrate		+	
	Ethyl isobutyrate	+	+	
	Ethyl penanoate	+	+	
	Ethyl 2-methylbutyrate	+	+	
	Ethyl 3-methylbutyrate	+	+	
	Ethyl hexanoate	+	+	+
	Propyl hexanoate		+	
	Isopentyl hexanoate		+	
	Ethyl heptanoate	+	+	

Table III (continued)

	R	C	W
Methyl octanoate		+	
Ethyl octanoate	+	+	+
Propyl octanoate		+	
Isobutyl octanoate		+	
Isopentyl octanoate		+	
Ethyl nonanoate		+	
Methyl decanoate		+	
Ethyl decanoate	+	+	+
Propyl decanoate		+	
Isobutyl decanoate		+	
Isopentyl decanoate		+	
Ethyl decanoate		+	+
Diethyl succinate		+	
Ethyl furoate		+	
Ethyl lactate		+	
Isopentyl lactate		+	
ETHERS			
Ethyl vinyl ether	+	+	
Propyl vinyl ether	+		
2-Methylfuran	+		
2-Vinylfuran	+		
2,5-Dimethylfuran	+		
Ethyl furfuryl ether	+		
HYDROCARBONS			
Isoprene	+		
Ionene	+		
Limonene	+		
p-Cymene	+	+	
KETONES			
Acetone	+	+	+
Butanone		+	
2-Pentanone		+	
3-Methyl-2-butanone		+	
2-Hexanone		+	
2-Heptanone		+	
2-Nonanone	+		
Diacetyl	+		
2,3-Pentadedione	+		
Acetoine	+		
2-Acetylfuran	+		
2-Methyl-3-tetrahydrofuranone	+		
Acetophenone		+	
SULFUR COMPOUNDS			
Methyl mercaptane	+		
Ethyl mercaptane	+		
Dimethyl sulfide	+		+
Dimethyl disulfide	+		+

The analyses were performed by GC-MS. Some compounds required additional verification by IR and NMR spectroscopy. Of the acetals diethoxyethane is quantitatively prominent in the beverages. Rum contained 83 ppm, cognac

30 ppm and whisky 25 ppm of this acetal, as could be determined by direct headspace analysis. As comparison, diethoxyisobutane occurred in a concentration of 1-6 ppm in the beverages.

Also the corresponding aldehydes, except formaldehyde, have been found in the headspace. Of these compounds ethanal is quantitatively the most important one. Many of the carbonyls are of value for the typical beverage flavour. Also the six ethers detected contribute to a certain extent to the top flavour of the beverages in which they were found. It is well known that esters belong to the important constituents of alcoholic beverages (4). The highly volatile esters could better be detected in the vapour phase than in the beverage itself. Of these esters ethyl acetate dominates by far. Second in concentration was ethyl formate.

Headspace enrichment of volatiles of essential oils was demonstrated by two gas chromatograms presented in Fig. 10. The upper chromatogram was obtained for a Russian clary sage oil, the lower one shows how the volatiles were enriched by application of the concentration technique, as illustrated in Fig. 9. The peak of linalool is marked on both chromatograms.

Table IV lists the compounds found in the headspace of clary sage oil.

Table IV. Headspace components of clary sage oil.

ALDEHYDES		ETHERS	
	Ethanal		2-Methylfuran
	Propanal		3-Methylfuran
	Isobutanal		2-Ethylfuran
	2-Methyl-2-propen-1-al	ESTERS	
	Pentanal		Ethyl formate
	2-Methylbutanal		Methyl acetate
	3-Methylbutanal		Ethyl acetate
	Hexanal		Hexyl acetate
	E-2-hexenal	HYDROCARBONS	
ALCOHOLS			Isoprene
	Methanol		2-Methyl-1-butene
	Ethanol		3-Methyl-1-butene
	Isopropanol	KETONES	
	Isobutanol		Acetone
	2-Butanol		3-Buten-2-one
	3-Methylbutanol		3-Pentanone
	2-Methyl-3-buten-2-ol		4-Methyl-2-pentanone
			3-Octanone
			Diacetyl

The components identified in the headspace of spike lavender oil are presented in Table V. Many of these components contribute significantly to

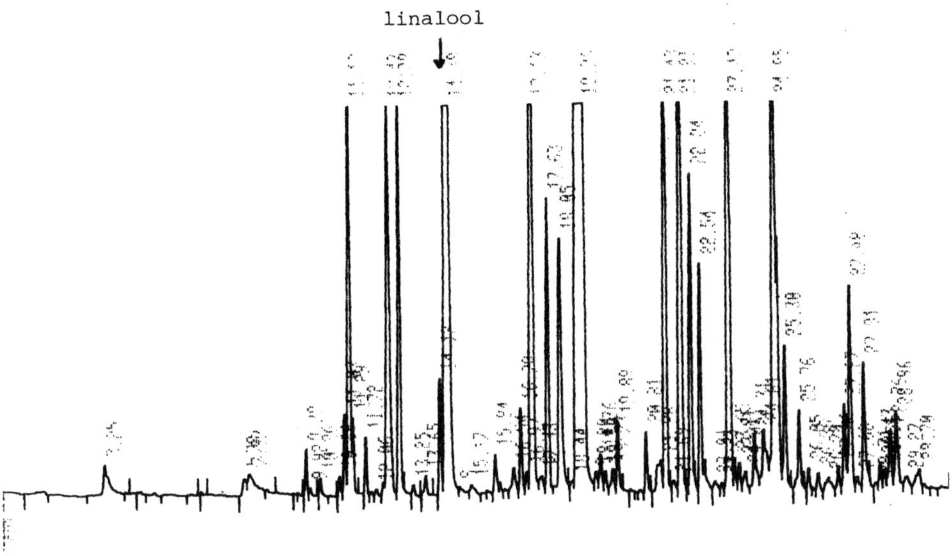

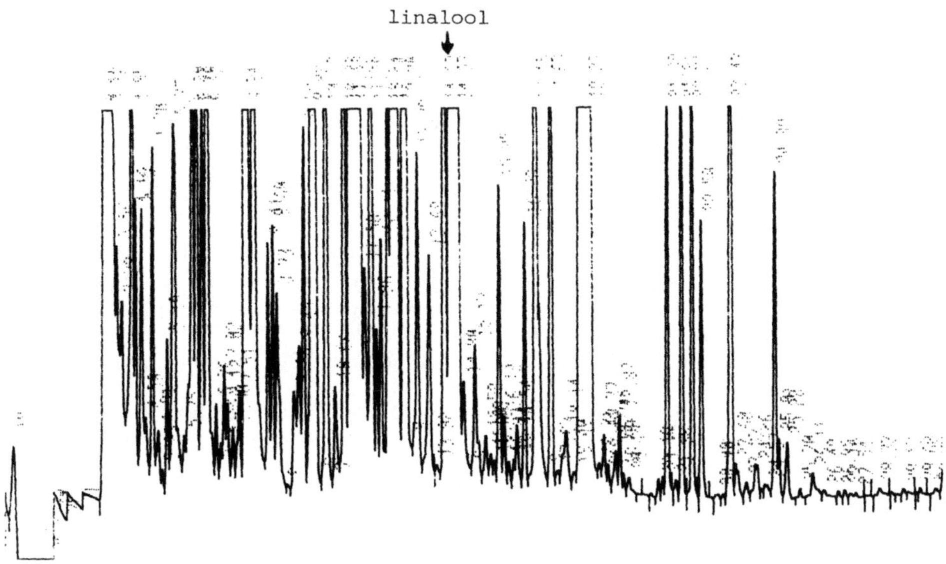

Fig. 10. Gas chromatograms of clary sage oil. Upper chromatogram, pure Russian clary sage oil; lower chromatogram, concentrated head-space of clary sage oil. Column, 50 m x 0.32 mm i.d., fused silica, df. 0.5 µm SE-54, temperature program, 60-250°C (4°C/min).

the top flavour of the oil.

Table V. Headspace components of Spanish spike lavender oil.

ALDEHYDES		ESTERS	
	Acetaldehyde		Isopropyl formate
	Butanal		Hexyl formate
	Isobutanal		Ethyl acetate
	Pentanal		Allyl acetate
	2-Methylbutanal		2-Butyl acetate
	3-Methylbutanal		Butyl acetate
	Hexanal		Hexyl acetate
	E-2-hexenal		Methyl butyrate
ALCOHOLS		HYDROCARBONS	
	Ethanol		Isoprene
	Isopropanol		Toluene
	Isobutanol	KETONES	
	2-Butanol		Acetone
	1-Penten-3-ol		Butanone
	2-Methyl-3-buten-2-ol		3-Buten-2-one
	Hexanol		2-Pentanone
ETHERS			3-Methyl-2-butanone
	2-Methylfuran		3-Hexanone
	3-Methylfuran		2-Methyl-3-pentanone
	2-Ethylfuran		4-Methyl-2-pentanone
	3-Isopropylfuran		2-Methyl-1-penten-3-one
	Hexyl methyl ether		3-Octanone

The methods discussed in this paper can be applied for concentration of the volatiles of different kinds of products. Excellent results have for instance been obtained with fruits, provided that the excess of water vapour is carefully removed.

REFERENCES

(1) Ter Heide R, de Valois PJ, Visser J, Jägers PP, Timmer R. 1978, in: Analysis of Food and Beverages (Charalambous G, Ed.), New York, Academic Press Inc., p. 249.
(2) Timmer R, Ter Heide R, de Valois PJ, Wobben HJ. 1971, J. Agric. Food Chem. 19:1066.
(3) Ter Heide R. 1983, in: Proc. Int. Symp. on Food Flavours (Adda J, Richard H, Eds.), Paris, Technique et Documentation - Lavoisier, APRIA, p. 27.
(4) Nykänen L, Suomalainen H. 1983, Handbuch der Aromaforschung, Aroma of Beer, Wine and Distilled Beverages, Berlin, Akademie-Verlag, p. 131.

SENSORY EVALUATION OF FLAVOUR AND FRAGRANCE COMPOUNDS AFTER HEADSPACE SAMPLING, CAPILLARY CHROMATOGRAPHY AND THERMAL CONDUCTIVITY DETECTION

H. OBBENS, L. HUBER

ABSTRACT

An analysis method is described, which gives a fast information on flavour and fragrance compounds of solid and liquid samples. The vapour phase above the thermostated sample is injected via an automated headspace sampling unit into a chromatographic system. The separation of the individual compounds is performed on a thickfilm fused silica capillary column with a high sample capacity. The separated compounds are detected by a thermal conductivity detector with a low filament cell volume and a high sensitivity. After the detection, the individual compounds can be used for further evaluation like sensory tests and spectroscopic analyses. The method does not require any sample preparation and can be used for liquid and solid samples.

INTRODUCTION

Gas chromatography is a well established method to characterize flavours and fragrances. The high resolution of capillary columns and the inertness of fused silica as tube material contribute to an excellent separation of the mixtures encountered in flavour and fragrances analysis.

Sensory evaluation of capillary peaks has been accomplished by splitting the column effluent between a flame ionization detector and a heated vent (1). The implementation of the thermal conductivity detector into capillary gas chromatography eliminated the need for effluent splitting (2).

The sampling of volatile components in solids or highly viscous samples can be done in different ways: 1. by extraction of the volatile components with a solvent, or 2. by direct sampling of the vapour phase by headspace analysis. A comparison of both methods has been made by Chialva et al. (3). Minimum or no sample preparation is the great advantage of the headspace sampling technique.

This paper describes the analysis of flavours and fragrances using headspace sampling, fused silica columns and thermal conductivity detection.

A. Baerheim Svendsen and J.J.C. Scheffer (eds), Essential oils and aromatic plants, ISBN-13: 978-94-009-5137-2

EXPERIMENTAL

Analyses were performed using a headspace sampling unit DANI HSS 3950 and a gas chromatograph Hewlett-Packard 5880A. The gas chromatograph was equipped with a split/splitless capillary inlet system, a thermal conductivity detector (TCD) and a flame ionization detector (FID).

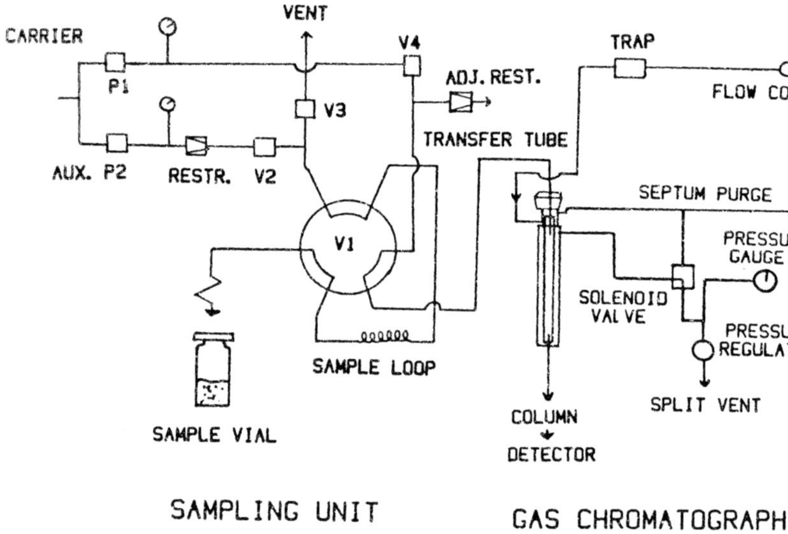

Fig.1. Scheme of the analysis system in the standby mode.

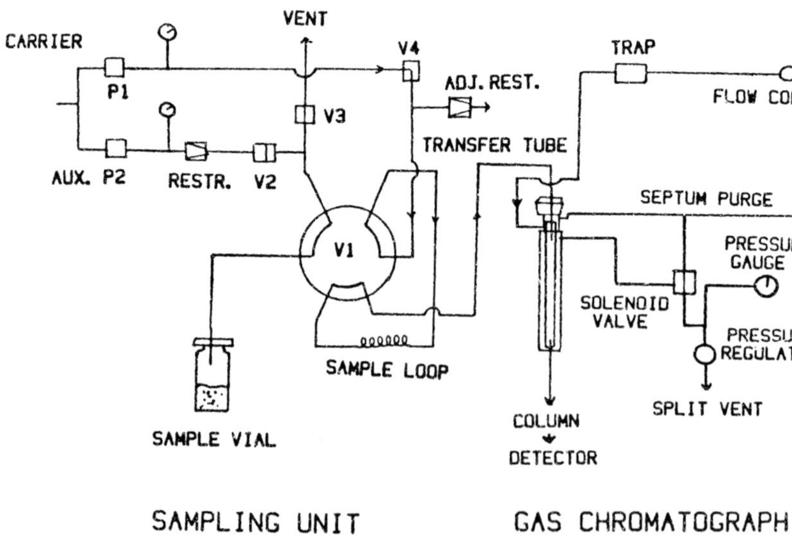

Fig.2. Scheme of the analysis system in the sampling mode.

The headspace sampler is a stand-alone unit, which can be interfaced to the gas chromatograph within a few minutes. A schematic diagram of the analysis system is shown in Fig.1. The main part of the sampling unit is a six-port rotary valve, which is used to realise the proper connections between the carrier gas, the sample loop and the GC injection port, the sample vial and an auxiliary gas. The size of the sample loop was 1 ml. The valve, the sample loop and the transfer line were heated up to $95^{o}C$ and the vial temperature was $80^{o}C$.

At the beginning of a sample cycle the sample needle moves down and penetrates the vapour space of the vial and the sample is pressurized by the auxiliary gas. After a preselected time of 10 s, valve 2 is closed and the loop equilibrates to the atmosphere via valve 3. After an equilibrium period of 10 s the valve 1 rotates and the content of the sample loop is flushed with the carrier gas into the injection port of the GC (Fig.2). At the same time, the capillary inlet system is switched from split into the splitless mode. 10 s later, all valves are switched back to the standby status (Fig.1). During this time, the transfer tube, the sample loop and the needle probe are flushed with Helium to prevent a contamination during the next chromatographic analysis.

The chromatographic conditions were: column, fused silica, 30 m x 0.53 mm, OV-1; oven temperature, $40^{o}C$ (5 min) $\rightarrow$ $180^{o}C$ ($6^{o}C$/min); detector temperature, $250^{o}C$.

RESULTS

The analysis of a fragrance sample was compared with a soap sample which was impregnated with the same fragrance. Fig.3 shows a chromatogram, which was obtained from the fragrance sample with TCD. A chromatogram obtained from a soap sample is shown in Fig.4. A comparison was made between TCD and FID. Figs. 5 and 6 show chromatograms obtained with an FID corresponding to chromatograms in Figs.3 and 4. The comparison shows that there is no difference in the peak shape obtained with the two detectors.

CONCLUSIONS

Headspace sampling is an ideal sampling method for solid and highly viscous samples. It eliminates the need for a sophisticated sample preparation. A stand-alone headspace sampler can be interfaced within a few minutes to different gas chromatographs. The splitless capillary inlet system enables the sampling onto capillary columns without a sample splitting. With this method the sensitivity of the analysis is increased. Large diameter capillary columns with thick films offer wide dynamic range and high capacity. The combination of these columns with single filament TCD provides a powerful tool for sensory evaluation of flavours and fragrances without the need of effluent splitting.

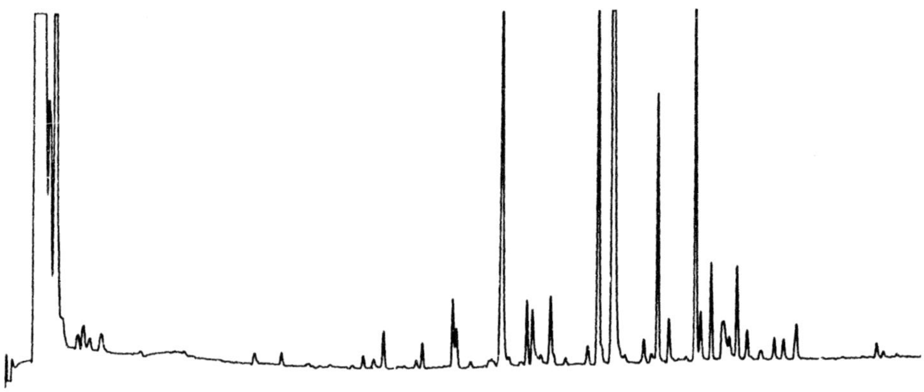

Fig.3. Fragrance sample with TCD.

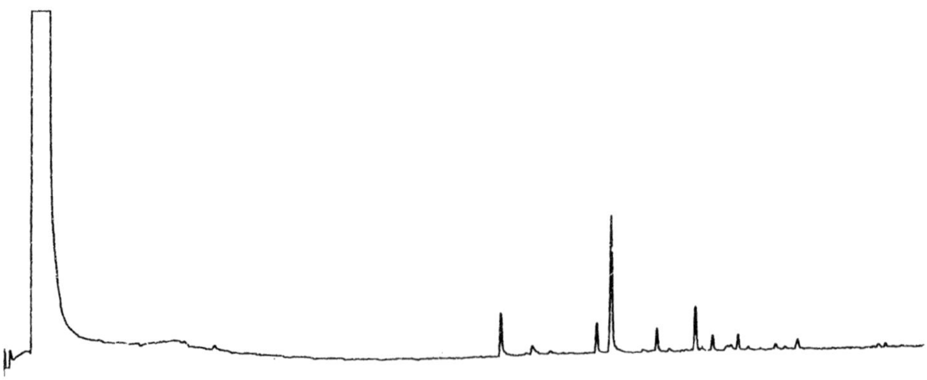

Fig.4. Soap sample with TCD.

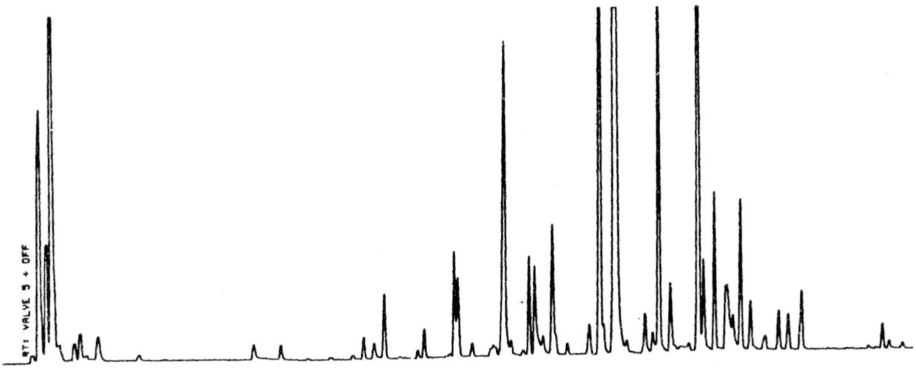

Fig.5. Fragrance sample with FID.

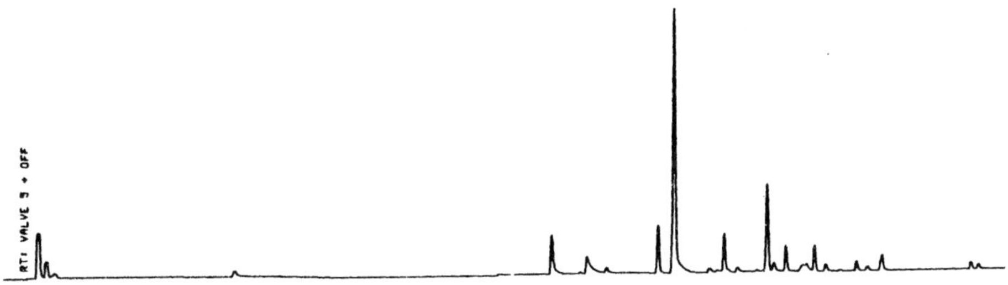

Fig.6. Soap sample with FID.

REFERENCES

(1) Sandra P, Talat Saeed, Redant G, Goodefroot M, Verstappe M, Verzele M. 1980, J. High Resol. Chromatogr. & Chromatogr. Commun. 3:17.

(2) Phillips R, Gratzfeld-Hüsgen A. Hewlett-Packard Application Note 228-38.

(3) Chialva F, Gabri G, Liddle PAP, Ulian F. 1982, J. High Resol. Chromatogr. & Chromatogr. Commun. 5:182.

GLC OF THE HEADSPACE AFTER CONCENTRATION ON TENAX GC AND OF THE ESSENTIAL OILS OF APPLES, FRESH CELERY, FRESH LOVAGE, HONEYSUCKLE AND GINGER POWDER

H.L. DE POOTER, B.A. COOLSAET, P.J. DIRINCK, N.M. SCHAMP

ABSTRACT

By concentrating the headspace components of intact fruits, vegetables and flowers on Tenax GC, followed by thermal desorption and GC-MS, the composition of the volatiles present can easily be studied. The technique may be applied to materials rich or poor in essential oils, yielding extensive information about the aroma compounds. Moreover, owing to the non-destructive character of the method, repeated sampling is possible. This makes it the method of choice for studying qualitative and quantitative changes in content and composition of the volatiles as function of time, e.g. in physiological and biochemical work.

INTRODUCTION

Until recently the flavour of aromatic herbs, spices, fruits, vegetables and flowers was usually assessed by analyzing the corresponding volatile oils, and only rarely by investigating the headspace. This is rather surprising since headspace analysis combines several worthwhile features: it gives a better image of the fragrance one actually observes by smelling, it is easily performed and may be repeated without trouble using different samples, and it is fundamentally non-destructive. Essential oils on the other hand are prepared by hydrodistillation or by extraction, which yield the sum total of the volatiles and destroy the matrix of the material in the process.

The easiest way of performing headspace analyses consists of direct sampling of the atmosphere above the material by means of a gas-tight syringe, followed by injection of the sample into a gas chromatograph equipped with a glass capillary column. This method gives good results with dried aromatic herbs, and complements the analyses of the corresponding essential oils to a large extent (1). However, it is less successful when sampling intact fruits, vegetables and flowers, because these release their volatiles more slowly and in smaller amounts into the atmosphere, so that only the most abundant components are detected (2-4). This difficulty can be overcome

A. Baerheim Svendsen and J.J.C. Scheffer (eds), Essential oils and aromatic plants, ISBN-13: 978-94-009-5137-2

by concentrating the headspace into enriched samples — for applications in the food industry, see (5) —, usually by adsorption on minute amounts of an adsorbent (ranging from carbon black, over GLC stationary phases, to Tenax GC), followed by elution of the compounds with an appropiate solvent.

In our laboratory, the volatiles are collected on relatively large amounts of Tenax GC and they are then thermally desorbed. The application of this method in fruit flavour analysis has recently been discussed (6). In the present paper its possibilities for studying the fragrance of fresh vegetables, flowers and fruits, and spices will be treated.

MATERIALS AND METHODS

Celery (*Apium graveolens* cv *dulce*), ginger powder (*Zingiber officinale*) and apples (*Malus domestica* cv Golden Delicious) were purchased locally. Lovage (*Levisticum officinale*) was obtained from the Botanical Garden, State University of Ghent, Belgium, and honeysuckle (*Lonicera periclymenum*) was collected in the woods of Ursel, Belgium.

Essential oils were prepared by hydrodistillation, followed by exhaustive extraction of the distillate with dichloromethane, and careful concentration.

Headspace collection was performed by flushing a glass sampling vessel (an 8 l desiccator for 1.5-2 kg intact apples, 90 g celery, 10 g lovage, 50 g honeysuckle, or a 250 ml roundbottomed flask for 1.5 g ginger powder), thermostated at 17°C, with air from a compressed air cylinder at 150-400 ml/min. After 15 min a Tenax GC 60-80 mesh adsorption tube (11 cm x 12 mm i.d.; 1.8 g adsorbent) was attached to the outlet of the glass container, and the sample was collected during 15-20 min. When closed tightly by means of glass stoppers and Parafilm M (American Can Cy), the loaded adsorption tubes could be stored at ambient temperature for long periods without loss or deterioration of the samples.

Treatment of apples with pentanoic acid: the acid (100 μl) was brought directly on the bottom of the desiccator, and the sweeping with air was continued (ensuring distribution of the vapours, and keeping the available amounts of acid within physiologically acceptable limits). Samples were taken 24 h after addition of the acid precursor.

Injection of the sample: for recovery and injection of the headspace concentrates, modified gas chromatographs equipped with a self-constructed desorption-injection system (Fig. 1) were used (6-10). The compounds were desorbed by heating the adsorption tube gradually to 180°C in a separate

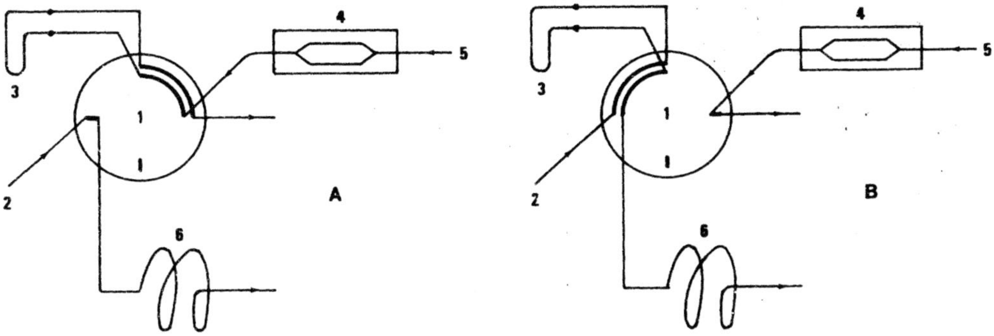

Fig. 1. Desorption-injection system: A. desorption and collection of the volatiles in a cold trap; B. injection by heating the cold trap. 1. two-position, six-port, high temperature injection valve (Valco Instruments, Houston, U.S.A.); 2. carrier gas, Helium, 4 ml/min; 3. cold trap (in liquid air); 4. desorption oven and adsorption tube; 5. sweeping gas, Helium, 10-12 ml/min; 6. glass capillary column.

oven under a continuous flow of Helium (10-12 ml/min) and carried away into a stainless steel collection tube (1/8" i.d.) cooled in liquid air (Fig. 1A). After completing the desorption and collection (30-45 min), the valve was turned into its second position (Fig. 1B), putting the cold trap in series with the glass capillary GLC column. The sample was then injected by heating the collection coil by means of a 1250 W flood-light or an oil bath at 180-200°C.

The corresponding essential oils were dissolved in carbon disulfide (1 μl in 100-250 μl of solvent) and 0.5-1 μl samples were injected.

Gas chromatography and mass spectrometry: the compounds were separated in Varian 2700 or 3700 instruments equipped with glass capillary OV-1 (40 m x 0.5 mm i.d.; coating thickness 0.75 μ) or SE-52 columns (150 m x 0.5 mm i.d.; coating thickness 0.6 μ; analysis of the apple headspace esters, Figs. 6A and 6B). Operating conditions: linear temperature program 10-200°C, 3°C/min (OV-1 column; Varian 3700 instrument) or 1°C/min (SE-52 column; Varian 2700 instrument); injectors, 180°C; detectors, 250°C; carrier gas, Helium, 4 ml/min.

For GC-MS, the glass capillary column, installed in a Varian 2700 gas chromatograph, was connected with the inlet of a MAT 112 mass spectrometer by means of a splitter and a platinum capillary (50 cm x 0.133 mm i.d.; 200°C).

Identification of the compounds: they were identified by their retention times as expressed by the Kováts indices using the homologous series of n-alkanes C_6-C_{18} as standards (11), by their mass spectra and, if possible, by comparison with authentic samples, which were synthesized or isolated, or present in essential oils of known composition (12).

RESULTS AND DISCUSSION

When comparing the GC traces of the headspace and the essential oil of the various samples used in this investigation, a clear difference was observed between the results obtained for fresh material (celery, lovage, honeysuckle and apples) and for dried ginger powder. Whereas the former tended to have more peaks with smaller Kováts indices for the headspace analysis, the latter gave analogous patterns for the headspace and for the essential oil.

As shown in Fig. 2, 13 components were immediately identified in the headspace of ginger powder, versus ten in the oil. Moreover, when larger samples of the headspace were examined, 16 more compounds were identified, namely α-thujene, α-pinene, camphene, sabinene, β-pinene, myrcene, Δ3-carene, α-terpinene, p-cymene, trans-β-ocimene, γ-terpinene, terpinolene, camphor, δ-elemene, a santalene and α-humulene, which compares favourably with published data for ginger oil (13, and references therein) up to β-sesquiphellandrene. Substances with a higher retention index, however, could not be detected.

The same phenomenon was even more pronounced for celery and lovage (Figs. 3 and 4), where the headspace seemed to be composed almost exclusively of mono- and sesquiterpenes while it was almost completely free of phthalides – celery character compounds (14). This was thought to be due in part to the thermal instability of these substances (15) on one hand, and to their low concentration in the headspace on the other. On stepwise investigation of the desorption-injection procedure, however, adsorption and decomposition of the phthalides in the graphite injection valve was found to be another possible reason for the absence of the substances in the gas chromatograms (Figs. 3A and 4A). Indeed, when the injection valve 1 (Fig. 1A) was bypassed during sample collection in the cold trap 3 and during injection (Fig. 1B), good peaks were obtained for phthalides present in a celery essential oil sample applied to a Tenax GC adsorption tube. An authentic celery headspace sample on the other hand, worked up in the same way, gave only slight deformation of the base line for the phthalides, indicating their low concentration in

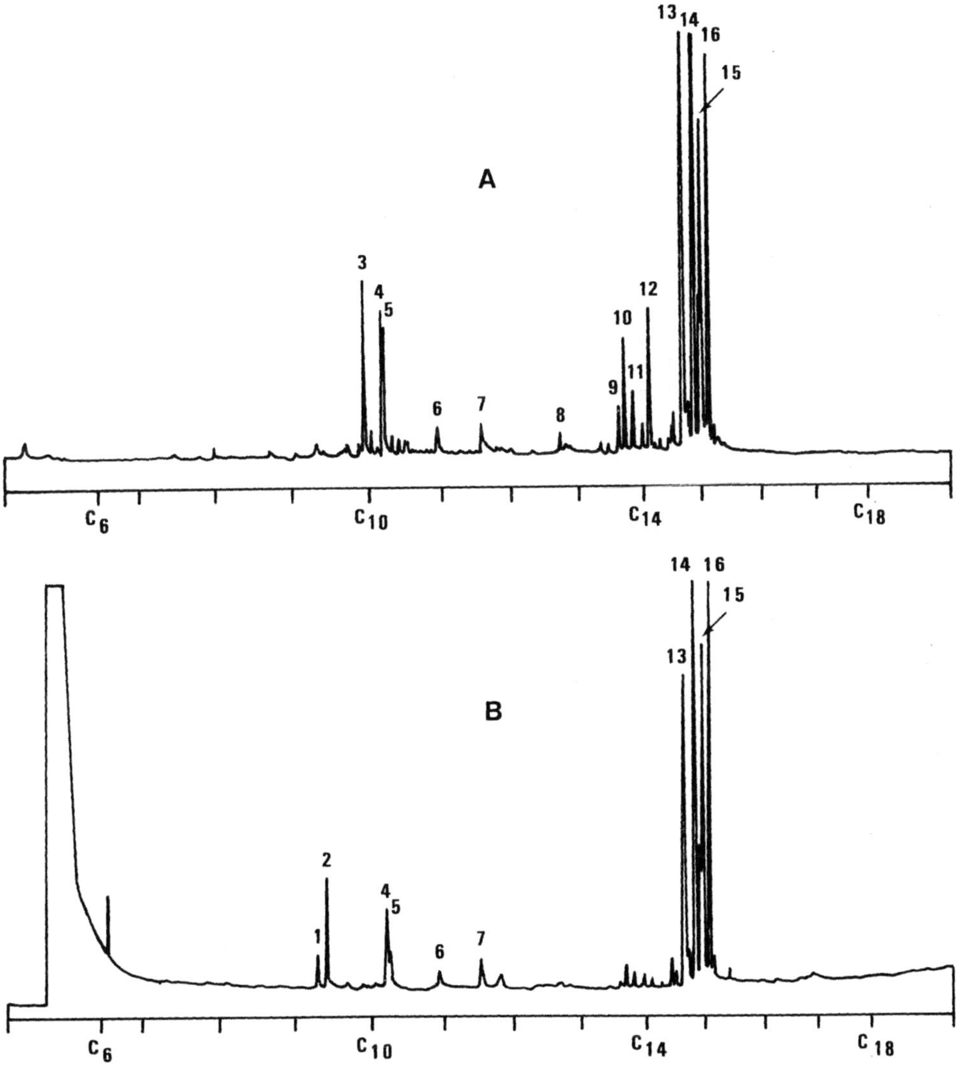

Fig. 2. Ginger powder (*Zingiber officinale*): A. headspace; B. essential oil; 40 m OV-1 glass capillary column.

1. α-Pinene
2. Camphene
3. α-Phellandrene
4. 1.8-Cineole
5. Limonene
6. Linalool
7. Borneol
8. Bornyl acetate
9. α-Copaene
10. Unidentified
11. β-Elemene
12. β-Caryophyllene
13. ar-Curcumene
14. Zingiberene
15. β-Bisabolene
16. β-Sesquiphellandrene

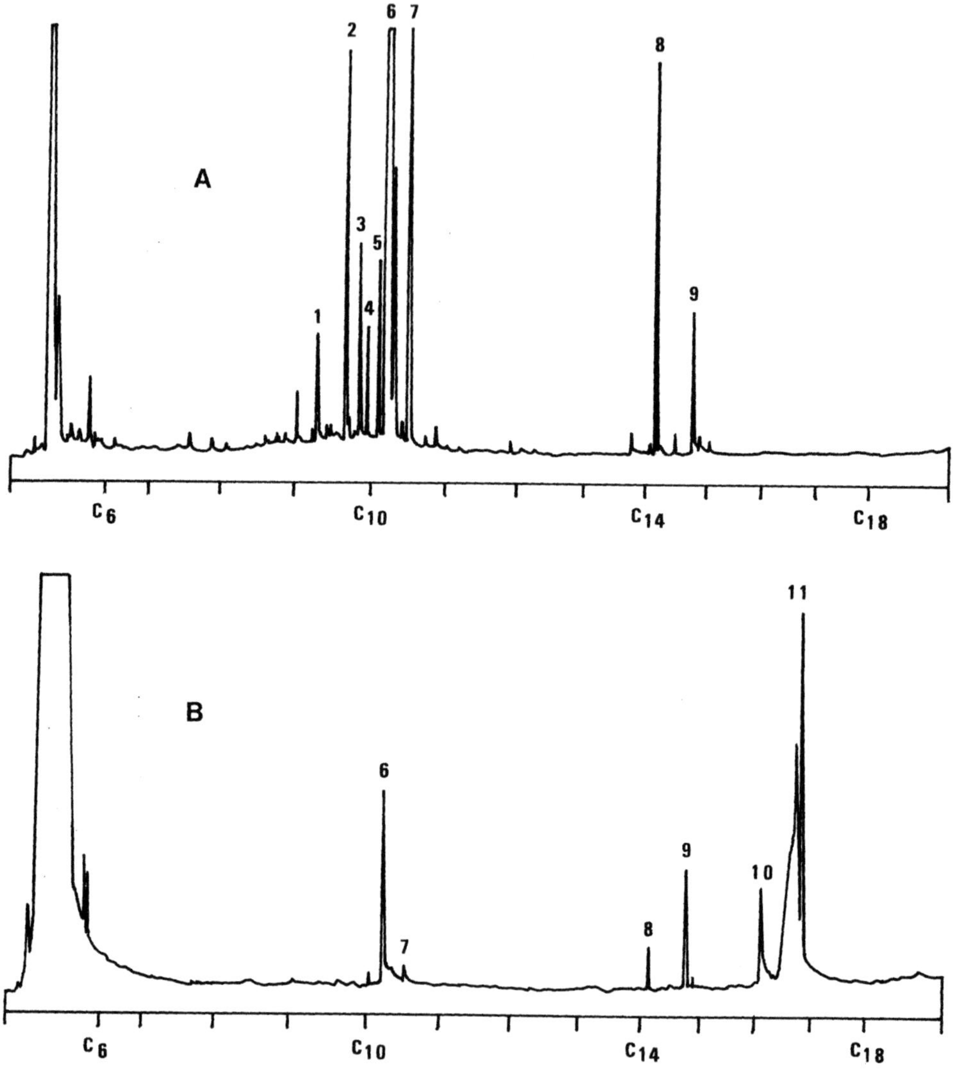

Fig. 3. Fresh celery (*Apium graveolens* cv *dulce*): A. headspace; B. essential oil; 40 m OV-1 glass capillary column.

1. α-Pinene
2. β-Pinene
3. Myrcene
4. α-Phellandrene
5. p-Cymene
6. Limonene
7. γ-Terpinene
8. β-Caryophyllene
9. β-Selinene
10. Butylphthalide
11. Mixture of hydrophthalides

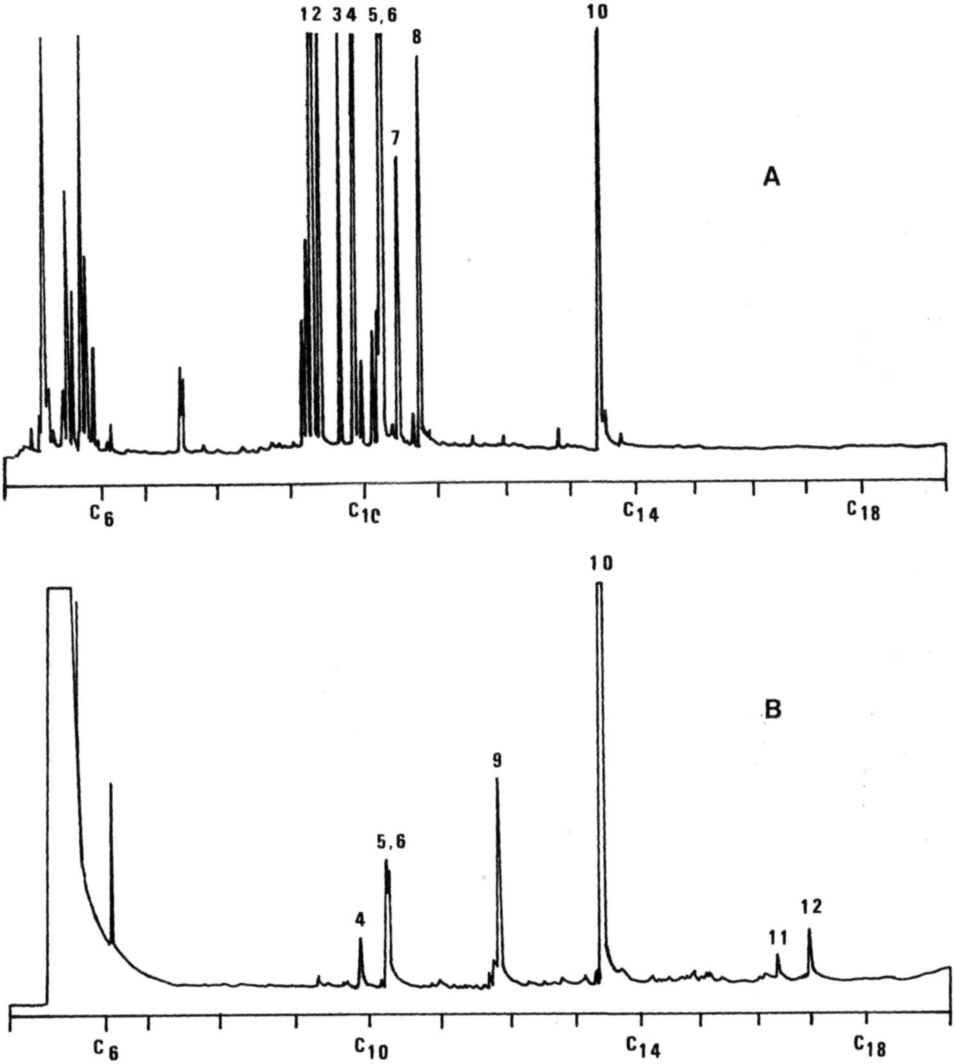

Fig. 4. Fresh lovage (*Levisticum officinale*): A. headspace; B. essential oil; 40 m OV-1 glass capillary column.

1. α-Pinene
2. Camphene
3. β-Pinene
4. Myrcene
5. 1.8-Cineole
6. Limonene
7. γ-Terpinene
8. Terpinolene
9. α-Terpineol
10. α-Terpinyl acetate
11. Butylidenephthalide
12. Ligustilide

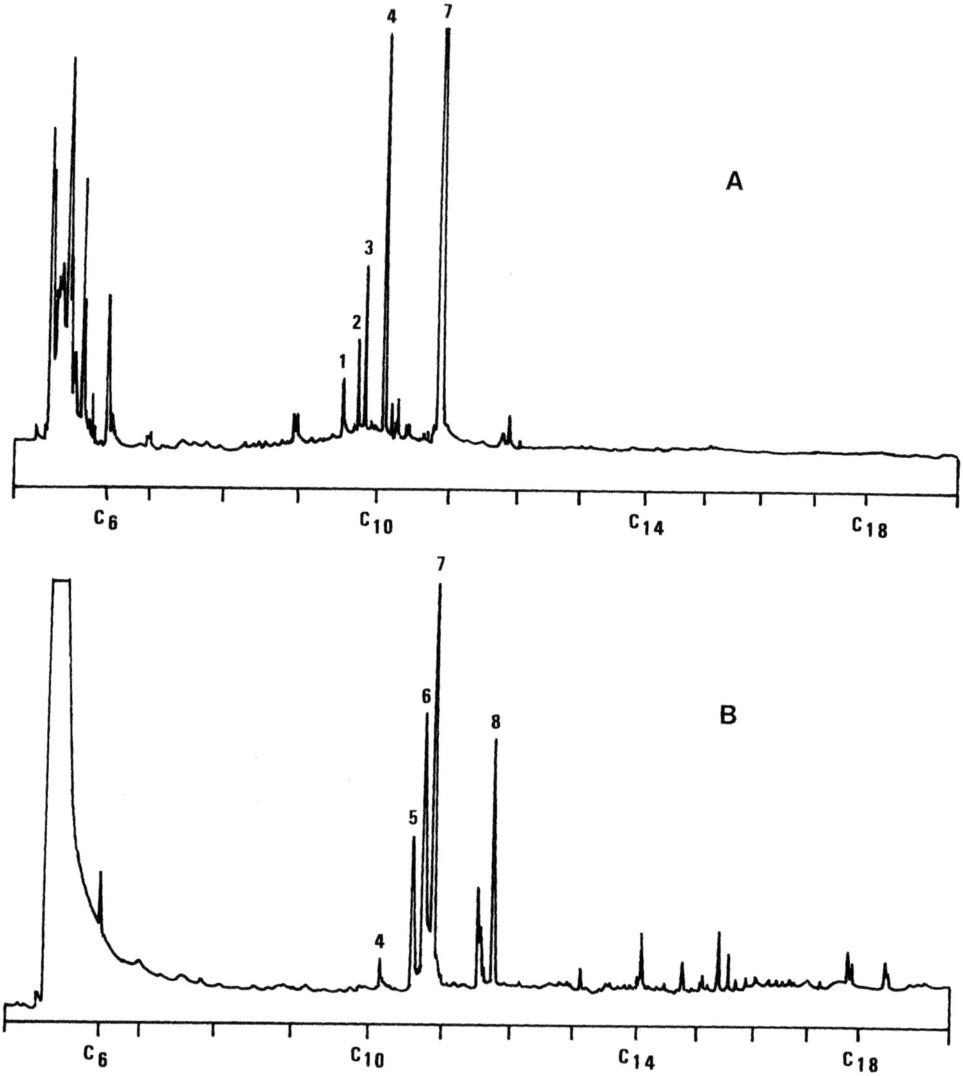

Fig. 5. Honeysuckle (*Lonicera periclymenum*): A. headspace; B. essential oil; 40 m OV-1 glass capillary column.

1. β-Pinene
2. Myrcene
3. α-Phellandrene
4. Limonene
5. Linalool oxide (trans ?)
6. Linalool oxide (cis ?)
7. Linalool
8. α-Terpineol

the headspace of intact celery.

The need for careful temperature control was confirmed while studying lovage headspace, the chief component of which (Fig. 4A) is α-terpinyl acetate (a tertiary alcohol ester and as such a member of a class of compounds which deacylate easily into alkenes). When the injection valve (Fig. 1) was heated to 250°C, and desorption and injection were performed at 220°C, no ester could be detected, but instead a very large limonene peak appeared. The situation returned to normal by reducing all temperatures to a maximum of 180°C.

As may be deduced from its headspace analysis, the odour of honeysuckle seems to be largely based on linalool and limonene (Fig. 5A). These components determine the overall fragrance impression, which is most probably refined and softened by minute amounts of other volatiles, the presence of which was shown by analysis of the essential oil (Fig. 5B).

An essential oil of Golden Delicious apples, prepared by hydrodistillation was notably poor in volatiles (mainly hexyl esters and α-farnesene; Fig. 6C), and — not surprisingly — had the odour of apple sauce. The headspace concentrate on the other hand was composed of a large variety of esters, mostly derived from acetic, butyric and hexanoic acids, and of α-farnesene (Fig. 6A).

The headspace enrichment technique described above thus allowed fast and relatively detailed evaluation of volatile contents up to substances with a Kováts index on OV-1 columns of 1500. Moreover, because sampling is performed without destruction of the materials, consecutive samples may be taken in function of time (in some cases spread out over a fortnight). As the method is quite reproducible (variation of the peak areas 10% or less), it was applied in a study of apple aroma formation, by addition of carboxylic acid precursors (9, 10). The changes in headspace composition of Golden Delicious apples after 24 h treatment with pentanoic acid vapours are illustrated in Fig. 6B. The formation of pentyl esters and pentanoates indicates the dependence of apple aroma formation on the availability of precursors, and not so much on the ester forming enzyme system (10).

CONCLUSIONS

By headspace enrichment on Tenax GC, followed by thermal desorption, the volatile content of intact vegetables, fruits, flowers, etc. may be determined. Since the sampling is non-destructive, the evolution of aromas may

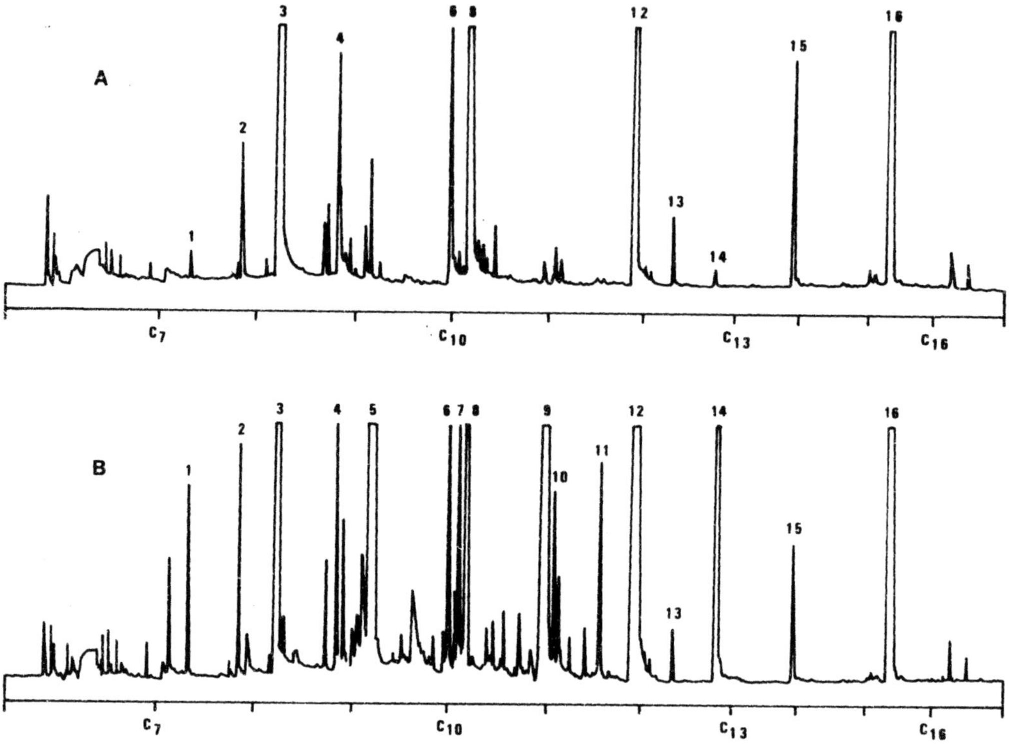

Fig. 6. Intact apples (*Malus domestica* cv Golden Delicious): A. headspace; 150 m SE-52 glass capillary column; B. headspace after 24 h incubation with pentanoic acid vapours at 17°C; 150 m SE-52 glass capillary column; C. essential oil; 40 m OV-1 glass capillary column.

1. Propyl acetate
2. 2-Methylpropyl acetate
3. Butyl acetate
4. 3-Methylbutyl acetate
5. Pentyl acetate
6. Butyl butyrate
7. Pentyl propionate
8. Hexyl acetate
9. Pentyl butyrate
10. Hexyl propionate
11. Pentyl 2-methylbutyrate
12. Butyl hexanoate + hexyl butyrate
13. Hexyl 2-methylbutyrate
14. Hexyl pentanoate
15. Hexyl hexanoate
16. α-Farnesene

be followed as function of time, e.g. in biochemical studies. As loaded adsorption tubes can be stored without alteration until GC analysis, many samples may be collected of materials which are only available during short periods of time, e.g. lilies of the valley in bloom. The method is only limited by the thermal stability and volatility of the compounds, limitations which are also found in GLC.

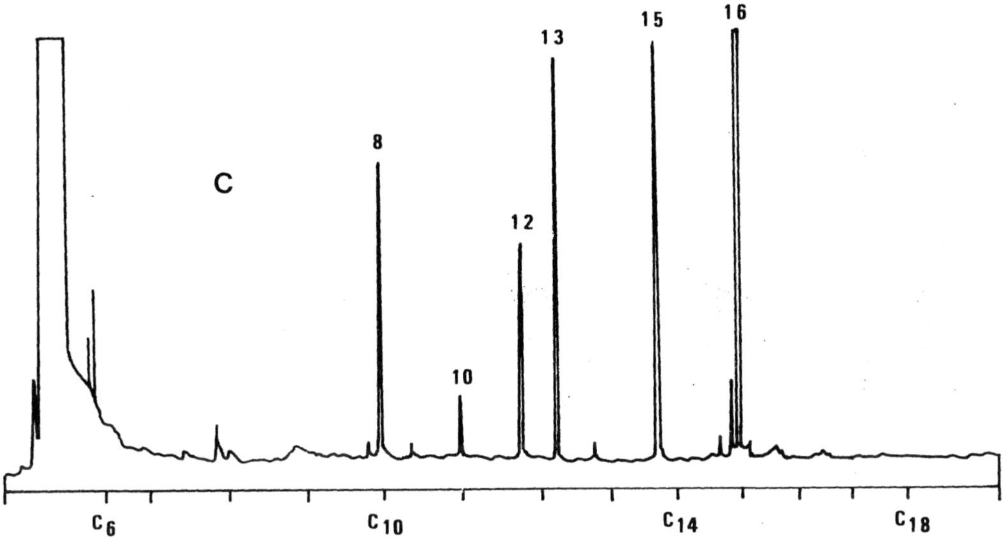

ACKNOWLEDGEMENTS

The 'Nationaal Fonds voor Wetenschappelijk Onderzoek', the 'Instituut tot Aanmoediging van het Wetenschappelijk Onderzoek in Nijverheid en Landbouw' and the 'Ministerie van Onderwijs' are thanked for their financial aid.

REFERENCES

(1) Chialva F, Gabri G, Liddle PAP, Ulian F. 1982, in: Aromatic Plants: Basic and Applied Aspects (Margaris N, Koedam A, Vokou D, Eds.), The Hague/Boston/London, Martinus Nijhoff Publishers, p. 183.
(2) Yamashita I, Iino K, Nemoto Y, Yoshikawa S. 1977, J. Agric. Food Chem. 25: 1165.
(3) Paillard N. 1979, Phytochemistry 18:1165.
(4) Knee M, Hatfield SGS. 1981, J. Sci. Food Agric. 32:593.
(5) Charalambous G. 1978, Analysis of Foods and Beverages. Headspace Techniques, London, Academic Press.
(6) Dirinck P, De Pooter H, Willaert G, Schamp N. 1984, in: Analysis of Volatiles (Schreier P, Ed.), Berlin, Walter de Gruyter, p. 381.
(7) Dirinck P, Schreyen L, Schamp N. 1977, J. Agric. Food Chem. 25:759.
(8) Dirinck P, De Pooter H, Willaert G, Schamp N. 1981, J. Agric. Food Chem. 29:316.
(9) De Pooter HL, Dirinck PJ, Willaert GA, Schamp NM. 1981, Phytochemistry 20:2135.
(10) De Pooter HL, Montens JP, Willaert GA, Dirinck PJ, Schamp NM. 1983, J. Agric. Food Chem. 31:813.
(11) Kováts E. 1965, Adv. Chromatogr. 1:229.
(12) De Pooter HL, Nor Omar M, Coolsaet BA, Schamp NM. 1985, Phytochemistry 24:93.
(13) Govindarajan VS. 1982, CRC Crit. Rev. Food Sci. Nutrit. 17:1.
(14) Wilson CW III. 1970, J. Food Sci. 35:766.
(15) Lund ED. 1978, J. Assoc. Off. Anal. Chem. 61:1083.

HEADSPACE TECHNIQUE AS A VERSATILE COMPLEMENTARY TOOL TO INCREASE KNOWLEDGE ABOUT CONSTITUENTS OF DOMESTIC OR EXOTIC FLOWERS AND FRUITS

D. LAMPARSKY

ABSTRACT

The so-called headspace technique has proved to be a versatile complementary tool to the classical analysis of all kinds of natural materials important to the fragrance and flavour industry.

Trapping of odorous compounds by this method followed by GC-MS analysis makes it possible to detect such constituents which may be lost during industrial treatment of flowers, leaves, fruits, etc. leading to extracts, absolutes or essential oils.

Furthermore, the headspace technique allows to set criteria for the genuineness of individual constituents and helps to explain the formation of artifacts. Since a headspace sample can be collected in a non-destructive way, it is also possible to apply this method to the investigation of time-depending changes in the production of volatile components within the living cells.

To demonstrate the great variety of possibilities in the fields of interest, some of our experiences with hyacinth (*Hyacinthus orientalis*), freesia (*Freesia refracta x armstrongii*), vanilla (*Vanilla planifolia*), lilac (*Syringa vulgaris*), honeysuckle (*Lonicera* species), verbena (*Verbena triphylla* or *Lippia citriodora* Kunth.) and strawberries (*Fragaria* spp.) will be discussed.

INTRODUCTION

Headspace methodology represents an important technique in analyzing natural substrates. The fragrance industry has already for a long time been aware of this, because what we are sniffing in our natural surroundings is always — in a large sense of the word — the headspace of flowers, leaves, woods, mushrooms, fruits and so on. That means that we perceive only those odorous molecules which are in the gaseous atmosphere around the plants.

The experimental basis of the method is nevertheless a closed system in which we find an equilibrium between the molecules in the vapour phase above a liquid phase and the ones in the liquid phase itself. The directing factor is the vapour pressure of the individual components. It is well known that the vapour pressure depends on the temperature of the system.

From this it becomes clear that low-boiling compounds having higher

A. Baerheim Svendsen and J.J.C. Scheffer (eds), Essential oils and aromatic plants, ISBN-13: 978-94-009-5137-2

vapour pressure, will be found in the 'headspace' to a larger extent than the high-boiling substances with only low vapour pressures. But we must keep in mind the general rule for a complex fragrance mixture: to have the true original smell, all components of the mixture should and will contribute to the overall sensation of a rose, a lilac, a hyacinth or others.

It is just this well-balanced mixture of a great variety of individual components which gives us the single impression we memorize as 'rose', 'lilac' or 'hyacinth'. Thus, mother nature is the best teacher for perfumers. If we can broaden our knowledge of the qualitative and quantitative composition of the really emanated scent of a highly fragrant flower and if we can link these data to the composition in the flower itself, we are a big step forward in the direction of the creation of fragrances 'closer-to-nature'. This is the practical background why our industry is interested in acquiring such knowledge.

We present in this paper some results illustrating the possibilities offered by headspace analysis alone and/or in combination with the analysis of a classical natural substrate such as an extract, an absolute or an essential oil, in which the formations of artifacts or losses of genuine components during the working-up procedures cannot be excluded.

HYACINTH

Just as reverence to the 'flowery Holland' we start with hyacinths, the springtime flowers with their typical fragrance. This is very pleasant in high dilution, but not always so in concentrated form due to its strongly earthy character. Our first flower is at the same time a good example for the great advantage of headspace techniques in detecting additional components never seen before in an industrially prepared absolute.

The absolute of hyacinth is produced in only small quantities; its gas chromatogram is shown in Fig. 1. We identified 115 constituents bearing the aromatic nucleus (1), that seems to be a very typical structural feature in the hyacinths. In addition, about 50 aliphatic compounds and only 30 terpenes were found. The group of terpenoids only occurs in low concentrations, i.e. cis- and trans-β-ocimene to the extent of about 0.005% and 0.02% respectively, and p-cymene as well as limonene as traces. These low percentages of terpenoids have raised the question if perhaps some other components of this type were lost during the industrial manufacturing. The best method to check this presumption is the headspace analysis.

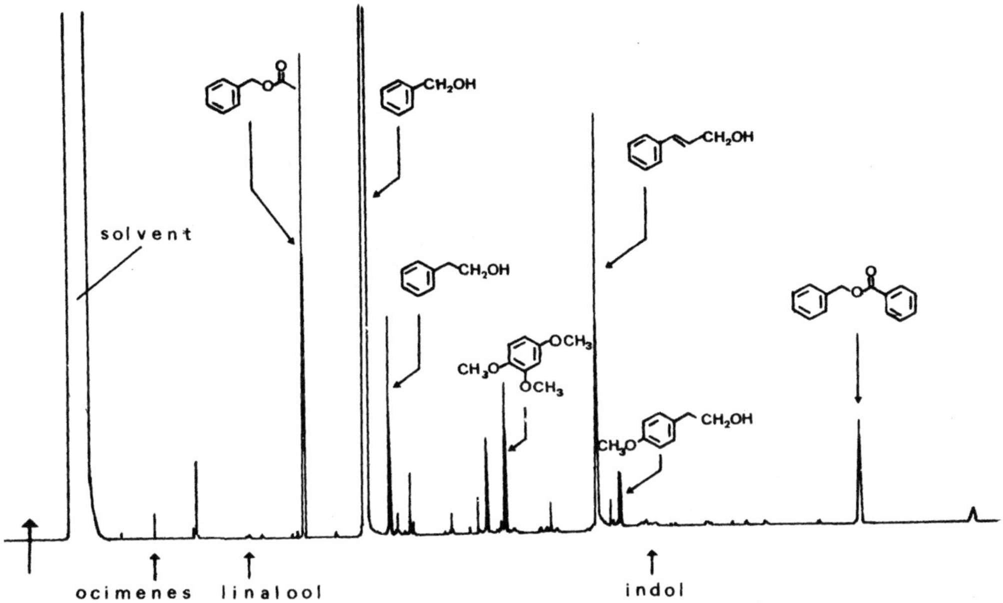

Fig. 1. Gas chromatogram of hyacinth absolute. Conditions: glass capillary column, 50 m x 0.3 mm, coated with UCON HB-5100; temperature program, $80^{\circ}C$ (6 min) then $80\text{-}180^{\circ}C$ ($3^{\circ}C$/min); carrier gas, Helium, 3 ml/min; injection, 1 µl of a 0.5% solution in pentane on cold column.

Fig. 2 shows a schematic view of the apparatus used for the experiments. Hyacinth flowers are placed on a grate. Nitrogen or other inert gases are passed through the system and take away all odorous molecules given off by the flowers. The emanating substances are subsequently adsorbed on small filters filled with charcoal or another adsorbent like Porapak. After a certain running period depending on the amount of volatiles emanated, and on the filter capacity (normally 100 µg/10 mg of charcoal) the trapped material is desorbed by means of a solvent like carbon disulfide, and analyzed via GC-MS. If the desorption is repeated at least three times, we never had serious difficulties as discussed by Curvers et al. (2). These authors recently pointed out that carbon disulfide is not very well suited for the desorption of alcohols and carbonyl compounds from charcoal. The GC fingerprint of the material collected via headspace sampling is shown in Fig. 3.

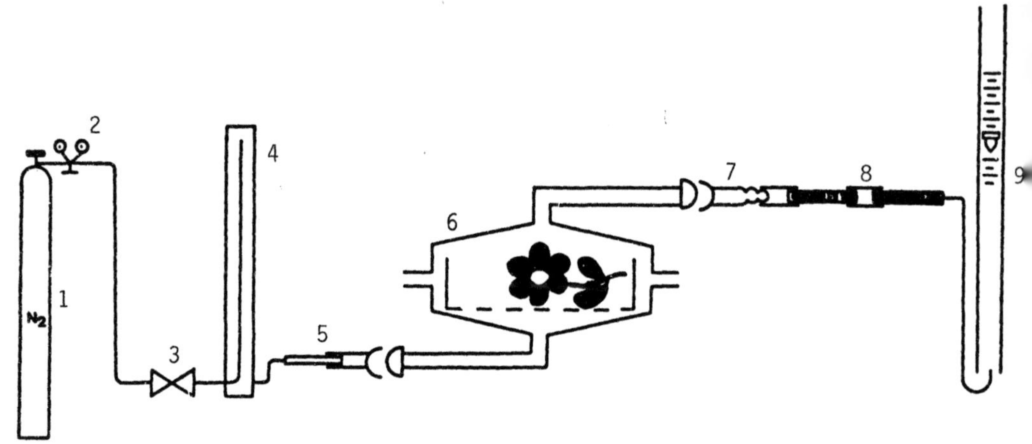

Fig. 2. Headspace apparatus (schematic). 1. Nitrogen cylinder; 2. reducing valve; 3. needle valve; 4. tower with molecular sieves; 5. connection; 6. vessel with grate; 7. connection; 8. charcoal filters (two); 9. rotameter.

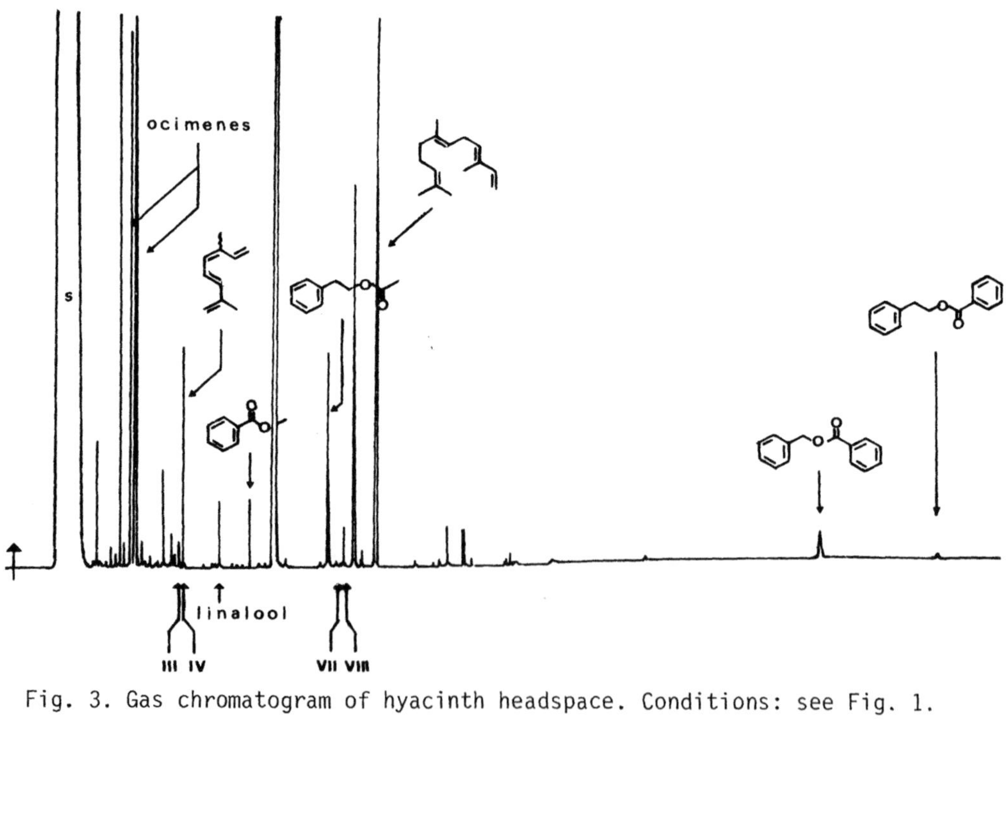

Fig. 3. Gas chromatogram of hyacinth headspace. Conditions: see Fig. 1.

As we already argued, the more volatile compounds are now present in higher concentrations. Furthermore, we could detect four new components marked with III, IV, VII, and VIII (Fig. 3). Their structures were confirmed by synthesis (Fig. 4).

Photo-oxygenation of ocimene(s) followed by reduction of the hydroperoxides formed leads to a mixture of four alcohols in the ratio of 42:3:31:24, the last two alcohols, 5 and 6, being the naturally occurring ocimenols. Treatment with thionyl chloride at $0^{o}C$ yields the two isomeric 2,6-dimethyl-octa-1,3,5,7-tetraenes, 3 and 4, the spectra of which correspond to those of the hydrocarbons found in the headspace sample of hyacinth flowers.

With respect to the ocimenols, 5 and 6, we have seen another interesting phenomenon, when analyzing the headspace of hyacinth flowers of different colours: these alcohols are present in the ratio of 1:2:3 in the blue, red and white blossoms respectively!

We are not sure about the biogenetic relationships which might be responsible for the differences observed, and we cannot exclude the possibility that only the degree of maturity of the blossoms will determine the concentration of these alcohols in the flowers. Meanwhile we have detected both ocimenols also in the headspace of lavender, *Osmanthus* blossoms and mango fruits as well as of various *Daphne* species exhibiting interesting odour profiles.

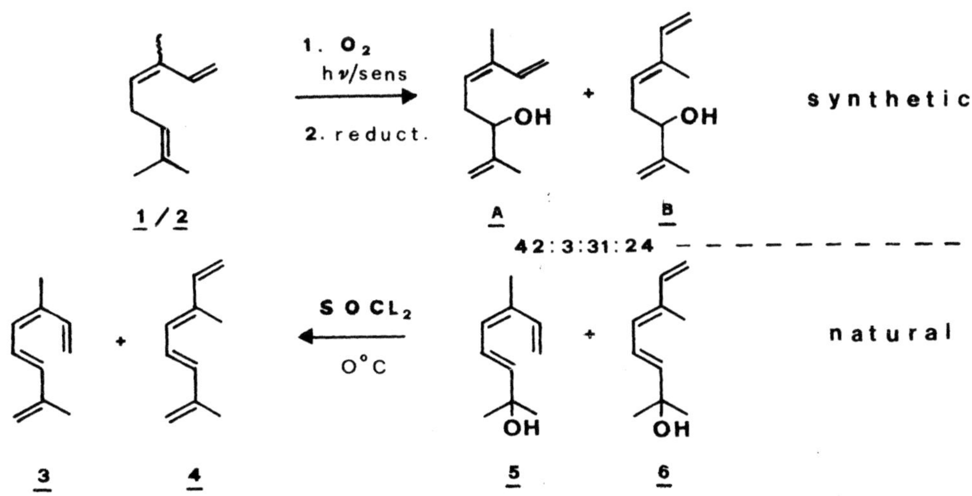

Fig. 4. Synthesis of monoterpenes newly detected in hyacinth flowers.

FREESIA

With respect to different quantitative compositions depending on the colour of the examined natural material, we have found another good example in the family of Freesia. Freesia flowers, nowadays available from flower shops in a great variety of colours, are hybrids of *Freesia refracta* and *Freesia armstrongii* originating from South Africa. Although the flowers are very popular and the emanated fragrance is delightful, nothing is known about their constituents.

Headspace samples were collected in the usual way by means of our laboratory apparatus. The flow rate was 40 ml/min of Nitrogen and two charcoal filters were placed at the exit. As expected, linalool was the main component. Together with α-terpineol it constituted 80% of the collected material. The ocimenols mentioned above could also be detected in the headspace of freesia, but only in that from yellow flowers. As in the case of hyacinth absolute, these tertiary alcohols were not present in extracts of

FREESIA

		white		yellow		red	
	%	LN	HS	LN	HS	LN	HS
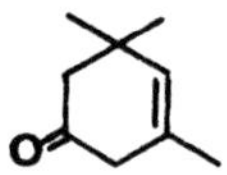	7	0.01	–	0.38	–	12.8	0.02
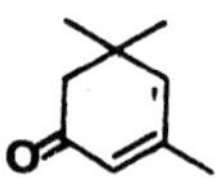	8	tr	–	0.05	–	3.8	tr
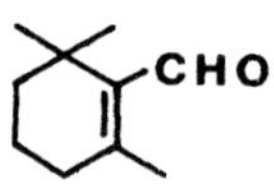	9	0.02	–	0.23	tr	0.08	–
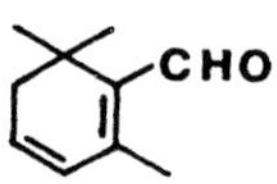	10	0.08	–	0.19	–	0.58	tr
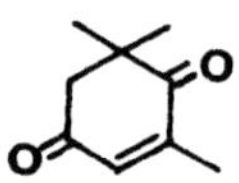	11	0.01	–	0.24	–	1.30	0.03

Fig. 5. Cycloaliphatic carbonyl compounds in freesia flowers of different colours. LN = Likens-Nickerson; HS = headspace; tr = trace.

freesia flowers obtained via a Likens-Nickerson apparatus.

The phenomenon of colour-depending composition was observed when the headspaces and extracts from white, yellow and red freesia flowers were analyzed (Fig. 5).

Only the extracts always contain β-isophorone, 7, α-isophorone, 8, β-cyclocitral, 9, safranal, 10, and oxoisophorone, 11. The concentrations increase with the deeper colour and we may assume that degradation of carotenoids plays a role in their formation. Only the red flowers exhale some of these components, thus indicating that in this case they are possibly genuine constituents. When sniffing at the red freesia, we immediately remark this contribution by a rather disagreeable vegetable- and isophorone-like odour note in comparison with the very clean, highly diffusive linalool-type odour of both the other light-coloured specimens.

Also some nitrogen-containing compounds were detected in the Likens-Nickerson extract of the red freesia flowers (Fig. 6). The alkylated pyridins, 12 and 13, and quinoleine derivatives, 14 and 15, could not be detected in the corresponding headspace material.

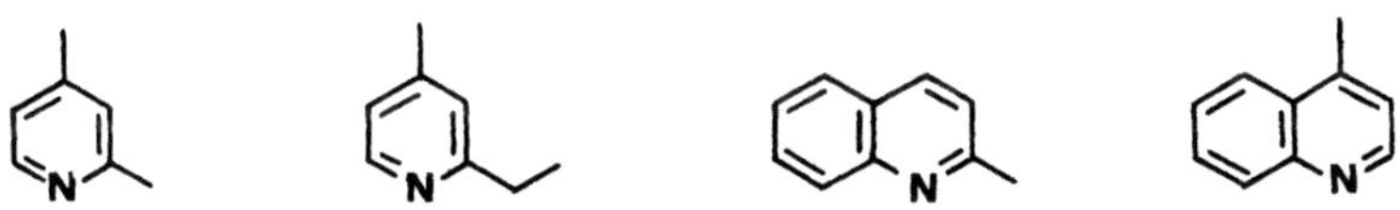

12 13 14 15

Fig. 6. Nitrogen-containing compounds in Likens-Nickerson extracts of red freesia flowers.

HONEYSUCKLE

The nitrogen-containing substances lead to another chapter in flower fragrance analysis. According to Exner (3) the fragrance of honeysuckle accounts for one of the most delicate odour types produced by nature. The extract of honeysuckle flowers does not correspond to the real fragrance of the flowers since a large number of ingredients are very sensitive to the isolation methods normally used.

However, the headspace method has yielded interesting informations concerning some nitrogen-containing compounds. As we have already reported (4), we were able to detect 2-methylbutyronitril, 16, as well as the syn- and anti-forms of 2-methylbutanal oxime, 17 and 18, in headspace samples of *Lonicera japonica* and *L. periclymenum*, the European variety of honeysuckle (Fig. 7).

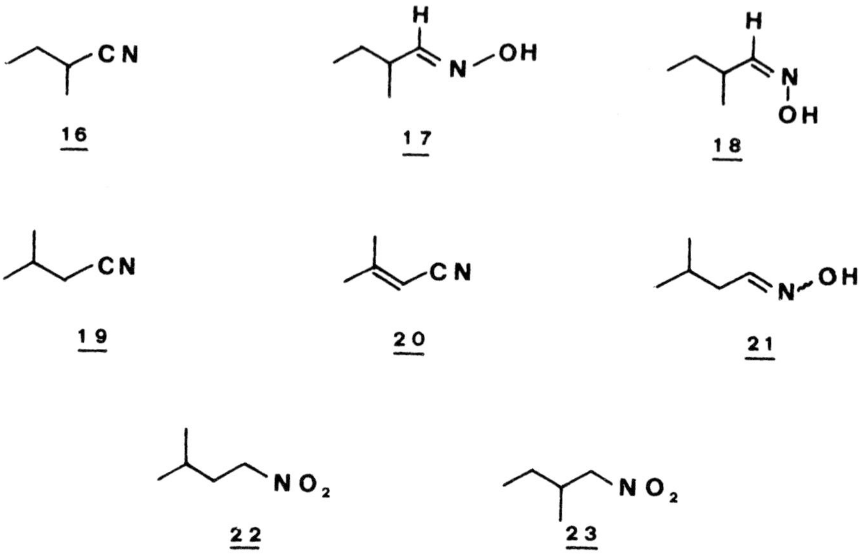

Fig. 7. Constituents of *Lonicera* species.

The first mentioned compound, 16, is also a component of black tea (5) and of orange flowers – *Citrus aurantium* ssp. *amara* – (4), the starting material for an absolute, well known for its perfumistic use. Due to its volatility, 16 could not be detected in the orange flower absolute of commercial origin.

A garden plant widely used in Europe is *Lonicera periclymenum* forma *serotina*. Its Likens-Nickerson extract contained about 0.1% 3-methylbutyronitril, 19, together with traces of the unsaturated analogue, 20. In the headspace sample compound 19 was present in higher concentration than in the extract. Surprisingly, the oxime of isovaleric aldehyde, 21, was another main component besides the compounds 17 and 18. The syn- and anti-forms of 21 constituted about 20% of the Likens-Nickerson extract!

The odour of *L. periclymenum* differs from that of *L. japonica*. It shows a distinct 'cinnamic shadow' which may be explained by the identification of 1-nitro-3-methylbutane, 22, accompanied by its isomer 23, 1-nitro-2-methylbutane. The ratio of these components was 20:1, and the forma *serotina* contained 10 times the amount of 22 compared with *L. periclymenum* itself.

There are not only interesting nitrogen-containing substances in the headspace material of various honeysuckle species. In the field of

LINALOOL OXIDES : japon. 10 : pericl. 1

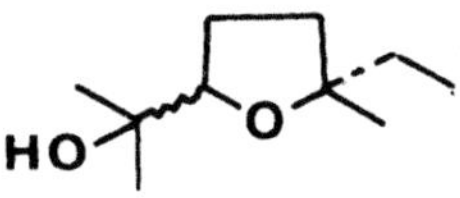

24 a

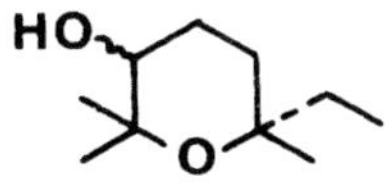

24 b

HS : 0.4 %

LN : 0.6

GERMACRENE-D

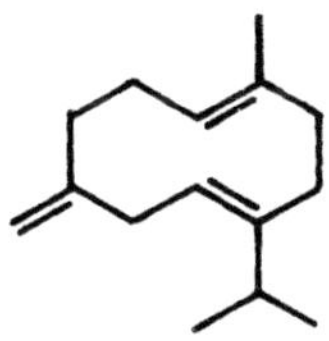

HS : 21.0 %

LN : 6.2

25

Fig. 8. Some terpenoid constituents of *Lonicera* species. japon. = *L. japonica*; pericl. = *L. periclymenum*; HS = headspace; LN = Likens-Nickerson.

terpenoids, for instance, we found in *L. japonica* 10 times the amount of linalool oxides, 24, (furanoid and pyranoid forms) in comparison with the European subspecies. Germacrene-D, 25, dominated by far all other sesquiterpene hydrocarbons, showing once again that it represents one of the most prominent genuine sesquiterpene hydrocarbons in the plant kingdom. Because of its sensitivity to acids and heat, germacrene-D (Fig. 8) will often not be found in industrially treated natural materials. It is then subject to further cyclization reactions giving a broad range of bi- and tricyclic derivatives, or a gum of polymeric material which cannot be isolated by steam distillation.

LILAC

When discussing analytical results of headspace samples of flowers, we should also mention the lilac (*Syringa vulgaris*) about which pioneering work was published by Mack and Köpsel (6). At the same time we worked at the same subject; a typical capillary gas chromatogram of a lilac headspace is shown in Fig. 9.

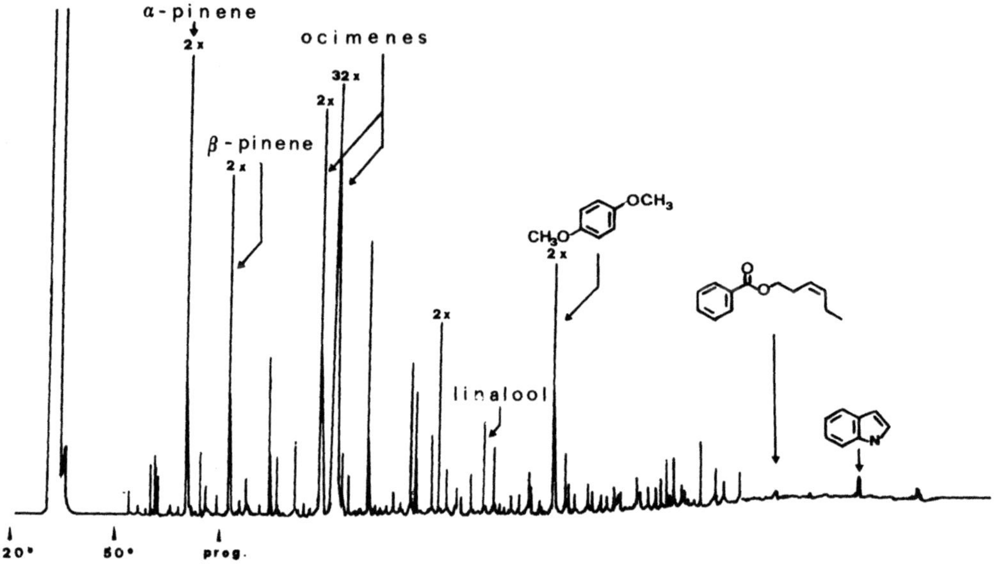

Fig. 9. Gas chromatogram of lilac headspace. Conditions: glass capillary column, 50 m x 0.33 mm, coated with UCON HB 5100; temperature program, 50°C (9 min) then 50-180°C (3°C/min); carrier gas, Helium; injection, 0.5 µl on cold column.

Besides the known components we identified camphene, limonene, camphor and β-bourbonene, and some additional aromatic compounds: p-methylanisole, phenylacetic aldehyde, estragole, p-methylacetophenone, 2-methoxybenzyl methyl ether, methyl eugenol, 1,2,4-trimethoxybenzene, piperonal, benzyl tiglate and cis-3-hexenyl benzoate.

Like in the case of hyacinth, the cis- and trans-ocimenes were the most prominent peaks in the headspace chromatogram. Mack and Köpsel (6) pointed out that these hydrocarbons may be formed by dehydration of linalool during the contact with charcoal in the trapping device. We checked this assumption by adsorbing the exhaled fragrant components on Porapak. We did not find any significant difference in the concentrations and the ratios of linalool and the ocimenes in both experiments. It seems therefore reasonable to accept that a large amount of the two hydrocarbons and a small amount of linalool occur really in the headspace.

This point is of great practical interest for the odour matching trials in our industry. The reconstitution of an odour 'true-to-nature' is not an easy task. We have to deduce the quantitative composition of the odorous mixture of a flower from data obtained by the headspace technique for components from the vapour phase. At the beginning it was a fastidious task of 'trial and error' before we were acquainted with the behaviour of all the fragrant compounds having big differences in their vapour pressures.

Therefore we prefer an additional quick analysis of an extract made by either the Likens-Nickerson method or the supercritical carbon dioxide extraction technique. Our collaborators Calame and Steiner (7) reported on such an extraction giving a lilac concrete (yield 0.0024%) which provided an analysable absolute oil (yield 0.00022%) after dewaxing. The main components of this absolute were: benzyl benzoate, 32.6%; elemicin, 14.6%; lilac alcohol (isomers), 9.8%; phytol, 7.8%; benzyl alcohol, 7.3%; cinnamic alcohol, 4.1%; hexadecanol, 3.2%. Ocimenes (0.01%) and linalool were only trace components, whereas cis-3-hexenyl benzoate was present to an extent of 0.5%, clearly showing the expected differences in the quantitative composition of a headspace material and the corresponding flower extract. It is interesting to note that Mack and Köpsel (6) found none of these main components except for the lilac alcohols.

VERBENA

As demonstrated above the deduction of useful data for practical reconstitution approaches will normally be facilitated by a parallel analysis of headspaces and extracts. This double effort may enlighten some other facts concerning, for instance, the presence or absence of Z- and E-isocitral in the verbena plant, *Lippia citriodora* Kunth.

Verbena oil obtained by steam distillation from the freshly harvested herb is actually forbidden for use in fragrance compositions, due to phototoxic effects. When analyzing verbena oil of the African- or the Provence-type, we found the photocitral isomers (8) besides a large amount of neral and geranial (together 40-60%). Especially the Provence-type oil contained the isocitrals, 26 and 27, (Fig. 10) – to an extent of more than 1%. These very instable aldehydes could be isolated by semipreparative GLC on packed columns. Interpretation of their spectral data led to the structures shown in Fig. 10.

Since the corresponding alcohols, 28 and 29, were also detected in the oil, the question is whether or not these compounds are genuine components. By GC analysis of the headspace of the flowering herb, we were not able to confirm the presence of the isocitral isomers. We therefore believe

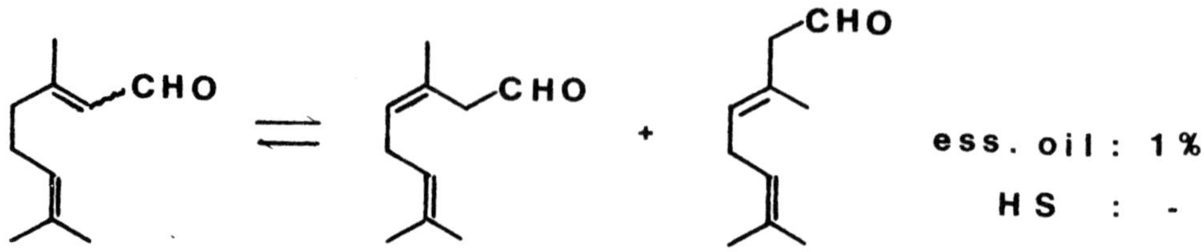

26 **27**

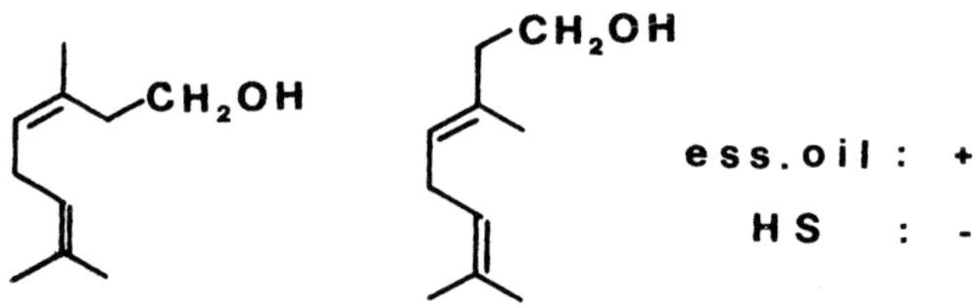

28 **29**

Fig. 10. Isomeric citral derivatives present in verbena oil and absent in headspace (HS) of flowering verbena herb.

that Z- and E-isocitral are artifacts formed during the production of verbena oil, as already supposed by Ohloff in 1960 (9). The heat-induced isomerization starts at temperatures reached in the manufacturing process.

VANILLA

Up to now we discussed fragrant components from flowers and herbs which were examples for the utility of a headspace analysis in combination with a 'classical' study of natural substrates.

The analysis of vanilla extracts obtained by solvent extraction of the beany fruits of *Vanilla planifolia* is somewhat difficult. The identification of trace components responsible for the better performance of a true 'natural vanillin' in vanilla flavours will be rendered troublesome because of the large amounts of vanillin, waxes and fats present. It was really our chance that the non-volatiles as well as the excessive amounts of vanillin, covering all other components in the GLC runs of the extracts, did not enter into the headspace sample (10, 11). Thus, the most volatile compounds could readily be enriched by the headspace sampling technique and afterwards analyzed without being disturbed by the main component vanillin (Fig. 11). In the GC-MS experiments we detected such 'low-boilers' as pentanal, β-cyclocitral, hexan-2-one, heptan-2-one, octan-2-one, amyl

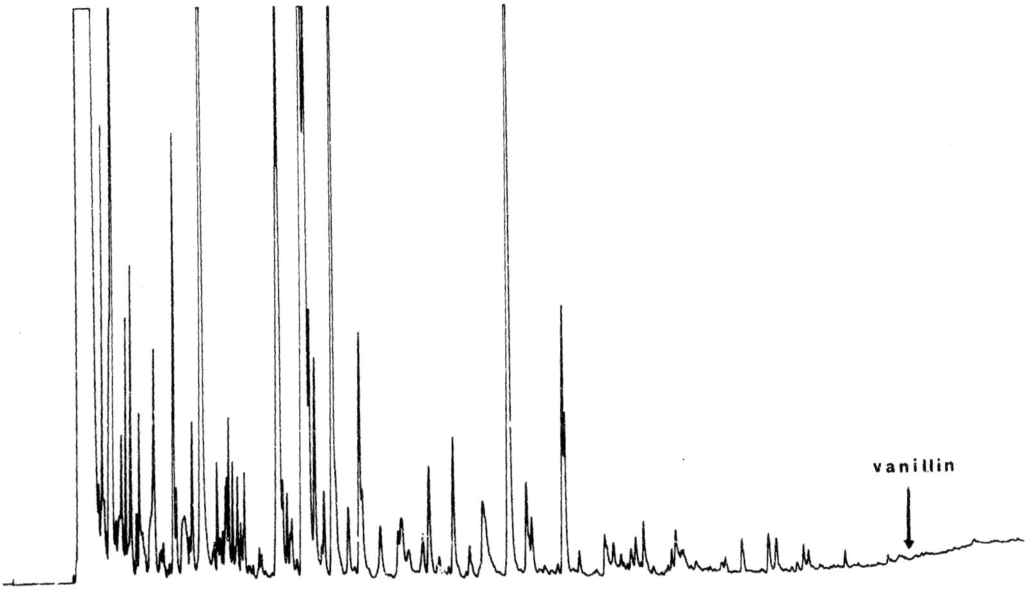

Fig. 11. Gas chromatogram of vanilla bean headspace. Conditions: glass capillary column, 50 m x 0.33 mm, coated with UCON HB 5100;temperature program, room temperature (15 min), 50°C (15 min) then 50-180°C (3°C/min); carrier gas, Helium; injection, 0.5 μl on cold column.

acetate, methyl valerate and others. Unfortunately, headspace techniques do not always offer the possibility to detect 'hidden peaks' when the main component — in this case vanillin — dominating all GC experiments, does not go at the same rate into the vapour phase above the natural product.

STRAWBERRY

Last but not least the flavour of strawberries is worth-mentioning to be discussed shortly. We all know the delicate, appetizing aroma given off by freshly picked fruits of the various *Fragaria* species. This fragrant material is influenced by the grade of ripeness and some day/night cycles of the constituents. We searched for a long time to find ways in which we could relate the most appreciated flavour to the chemical composition — at the best point of maturity (12).

Observation of the headspace composition of strawberries during four days led us to the following:

- some components decrease with the time, like methyl, ethyl, isopropyl and butyl hexanoates;

- some components increase with the time , like ethyl, isobutyl and isoamyl acetates as well as ethyl valerate – all signs of the start of fermentation within the fruits;
- some components show a distinct maximum, like propyl, isopropyl, isoamyl and hexyl butanoates;
- some components show a distinct minimum, like ethyl butyrate, methyl isovalerate and 2-hexenyl acetate;
- some components show an oscillating effect, like γ-nonalactone.

It was the tedious work of the flavourists to create the strawberry flavour using all data coming from their analytical colleagues. When the cell membranes of the strawberries are destroyed, after a short period reactions leading to a completely changed picture of the chemical composition are observed. So the composition does not show the correct balance of all important esters giving the typical touch of a really good strawberry flavour. Summarizing, the headspace techniques enable us to monitor the ripeness-depending changes in the composition of the volatiles from a single strawberry.

ACKNOWLEDGEMENTS

My co-workers and colleagues who have analytically worked with extraordinary enthusiasm in this fascinating field during the last years, Mr. W. Grab (strawberries), Mr. R. Kaiser (hyacinth, freesia, honeysuckle, verbena) and Dr. I. Klimes (lilac, vanilla beans, special methods), are most cordially thanked.

REFERENCES

(1) Kaiser R, Lamparsky D. 1977, Parfums Cosmet. Aromes No.17:71.
(2) Curvers J, Noy T, Cramers C, Rijks J. 1984, J. Chromatogr. 289:171.
(3) Exner F. 1984, Parfüm. Kosmet. 65:77.
(4) Kaiser R, Lamparsky D. 1982, in: Annales techniques 8. Congrès international des huiles essentielles, Cannes 1980, p. 287.
(5) Bondarovich HA, Giammarino AS, Reuner JA, Shephard FW, Shingler AJ, Gianturco MA. 1967, J. Agric. Food Chem. 15:36.
(6) Mack H, Köpsel M. 1973, Parfüm. Kosmet. 54:233.
(7) Calame JP, Steiner R. 1982, Chem. Ind. No.12:399.
(8) Kaiser R, Lamparsky D. 1976, Helv. Chim. Acta 59:1797.
(9) Ohloff G. 1960, Tetrahedron Letters No.11:10.
(10) Klimes I, Lamparsky D. 1976, Int. Flavours Food Addit. 7:272.
(11) Klimes I, Lamparsky D. 1978, in: Analysis of Food and Beverages (Charalambous G, Ed.), New York, Academic Press Inc., p. 95.
(12) Grab W. 1978, in: Proceedings International Federation of Fruit Juice Producers, Symposium on Flavours of Fruits and Fruit Juices, Bern, p. 213.

HANDLING, QUICK OPTIMIZATION AND CHOICE OF COLUMN WALL MATERIAL IN CAPILLARY GAS CHROMATOGRAPHY

W. GÜNTHER, F. SCHLEGELMILCH

ABSTRACT

The handling, i.e. installation of glass capillaries, is discussed. The modification of the underground pretreatments by straightening the glass capillaries is shown by SEM and the influence of this procedure is discussed. A very useful way of quick optimization for flame detectors is given. Also the fast setting up of carrier gas velocities is shown together with the possibility of calculating the velocity and the carrier gas flow. The aims in the choice of column wall materials are given. Soft glass, borosilicate glass, fused silica and quartz are discussed.

The installation only of glass and fused silica capillaries is discussed here, because today metal capillaries are seldom used. The installation of fused silica capillaries normally brings no problems because of their good flexibility. A too narrow bending ratio, however, should be avoided, because they can break. Alternatively this could cause a complete interruption of the capillary inside, the polyimide lacquer, because of the high brittleness of fused silica. It should be pointed out that also in the case of slitted polyimide or scratched polyimide there is the possibility of some carrier gas diffusion.

The connection of glass capillaries is principally not very difficult, if they are handled carefully. Just the end pieces must be straightened. The most used column wall material in Europe by now is glass. Therefore, this procedure will be given in the next figures. For straightening the glass capillaries normally only a common gas lighter is necessary. The straightening of borosilicate glass, however, is somewhat more difficult, because the softening range of borosilicate glass is about 80°C higher than that of soft glass. The ranges are for soft glass 500-650°C and for borosilicate glass 550-730°C.

These temperatures are the lowest to be used. The straightening of the glass capillary ends is to be done as shown in Fig. 1. As can be seen, the capillary must be held in the left hand and heated by a common gas lighter.

A. Baerheim Svendsen and J.J.C. Scheffer (eds), Essential oils and aromatic plants, ISBN-13: 978-94-009-5137-2

After having straightened the sagged end as shown in Fig. 2, the glass capillary is turned to a position as shown in Fig. 3, whereby a small zone of the straightened end undergoes further heat treatment. The correct temperature is reached when the heated zone becomes dark red; then the capillary is to be bent immediately in the desired direction with the second finger of the hand with the gas lighter.

The higher the flame temperature, the quicker the straightening can be done and the better the quality of the straightened glass capillary ends, because the pyrolytical decomposition of the stationary phase runs

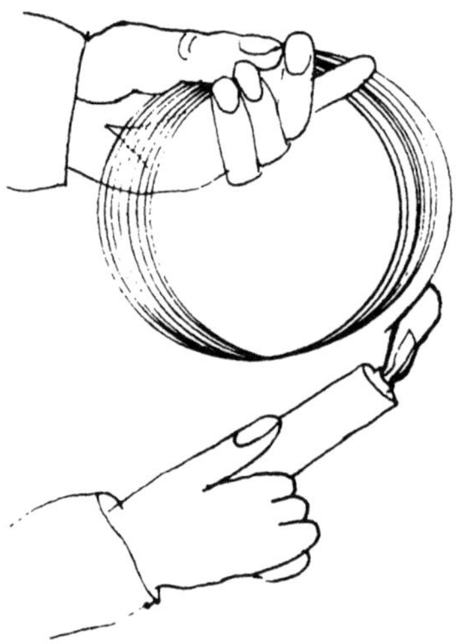

Fig. 1. Starting position for straightening glass capillary columns.

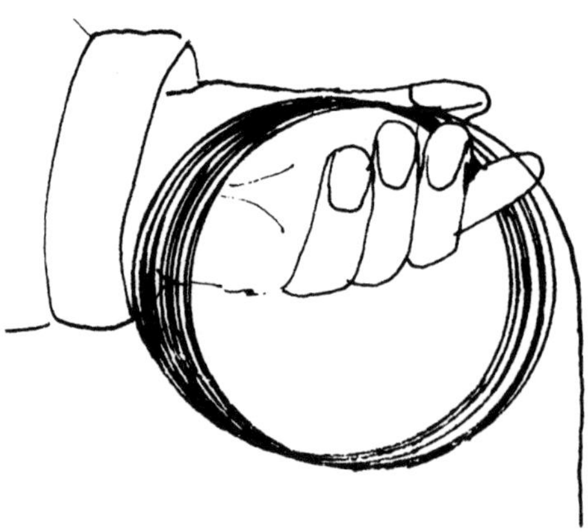

Fig. 2. Straightened glass capillary column end.

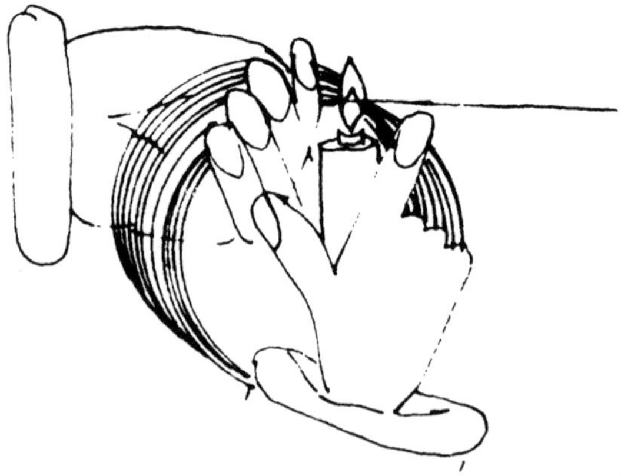

Fig. 3. Position for bending up glass capillary ends.

quantitatively down to carbon dioxide and water and similar components, the higher the straightening temperature is. Sometimes it is necessary to purge with Oxygen during the straightening procedure. This means that the use of electrical glass capillary straightening machines is not recommendable, especially when the manufacturer points out that these machines work about 400°C below flame temperatures. If the straightening temperature is too low, there is a formation of soot, as shown in Fig. 4.

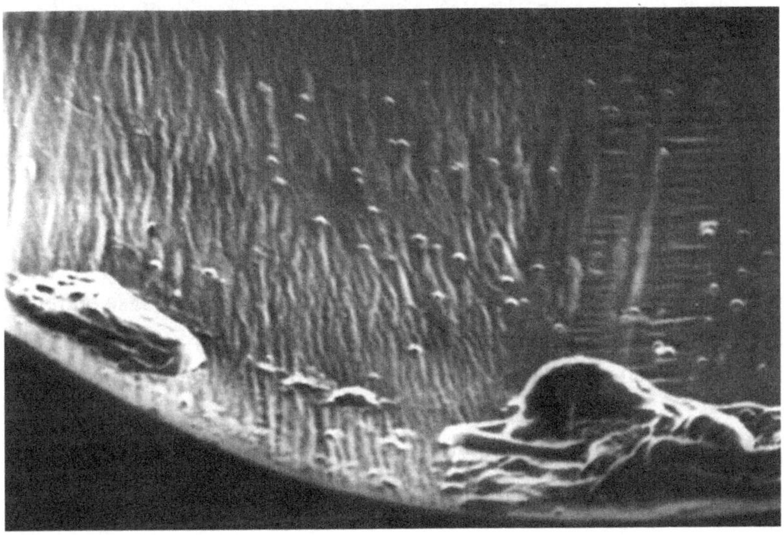

Fig. 4. Formation of soot in poorly straightened glass capillary ends.

It is easy to see that glass capillary columns with these kinds of straightened ends set up problems like decomposition of compounds to be analyzed or adsorption effects. On the other hand, well straightened glass capillary columns show well defined crystal undergrounds, as demonstrated in Fig. 5.

As can be seen, there is a sharp zone of well known NaCl crystals which are not reached by the high temperature of the gas lighter. Next to it is a dendritic NaCl formation arising from melted and recrystallized NaCl crystals. There is no formation of soot; this means that the stationary liquid phase was pyrolyzed quantitatively, and there is no residue.

The normally used gas flow in capillary GC is about 0.3-3 ml/min if Nitrogen is used. If Hydrogen is used the flow can be about ten times higher.

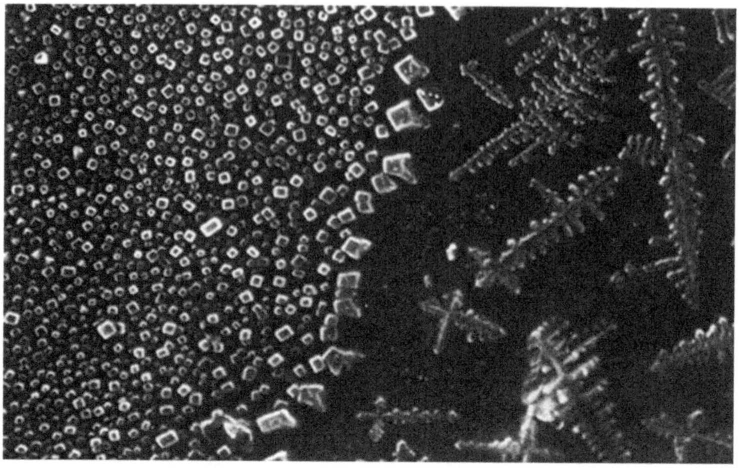

Fig. 5. Well straightened glass capillary with NaCl underground.

The flame detector gives no optimum results under these conditions. For this reason a make-up gas is absolutely necessary. A normal flame detector needs 30-50 ml/min. For purging the detector we, therefore, use a normal inlet splitting system. The split line is used for setting up carrier gas as make-up gas. This must be done by a flow controller. The results of giving make-up gas to the detector are given in Fig. 6.

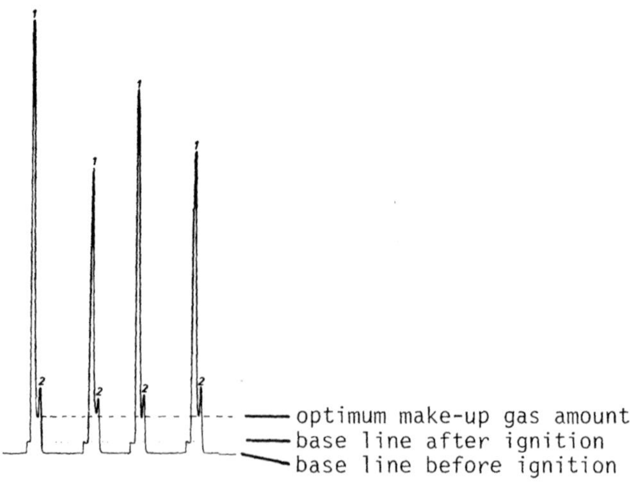

Fig. 6. FID response due to optimization with make-up gas.

After ignition the base line goes up somewhat. After purging with make-up gas two maxima will be seen. The optimum of make-up gas is pointed out just below point 2. More make-up gas will quench the flame. It is not necessary to know the exact amount of make-up gas, because this depends on the flow through the capillary column. This procedure needs about ten seconds. Just at this point the FID works with linear response factors and gives the highest sensitivity.

Another positive effect of the make-up system is the capturing of waste from the graphite and vespel ferrules used. Besides, there is no contamination of the FID by broken capillary column pieces.

In the case of changing the Nitrogen flow, a loss of more than 200 000 theoretical plates was observed for compounds such as alkanes (pentane - decane), benzene, ethanol, pyridine and nitromethane. With Helium as carrier gas the loss of theoretical plates was less than 50 000.

The differences can be explained in terms of the HETP-curves of the two carrier gases. There is no possibility for a quick setting up of the optimum carrier gas flow by using bubble flow meters. It should be strictly pointed out that there is absolutely no need for measuring gas flows instead of the linear gas velocity. For the optimization of GC columns it is necessary to know the linear velocity in order to calculate the HETP-minimum. In Fig. 7 the relation between the pressure and the linear gas velocity at different temperatures is shown.

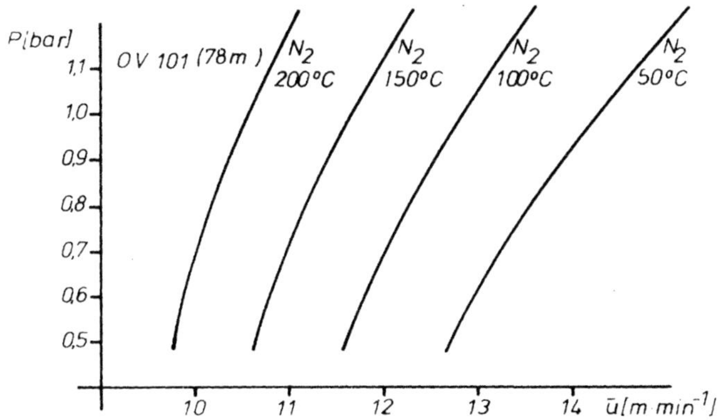

Fig. 7. Pressure and linear gas velocity at different temperatures.

The average linear gas velocity is easy to calculate as

$$\bar{u} = \text{column length} : \text{dead time}$$

If one must know the gas flow, the average linear gas velocity should be multiplied by πr^2. If $\bar{u}$ is calculated in dependence of the adjusted pressure the results are straight lines, as shown in Fig. 8.

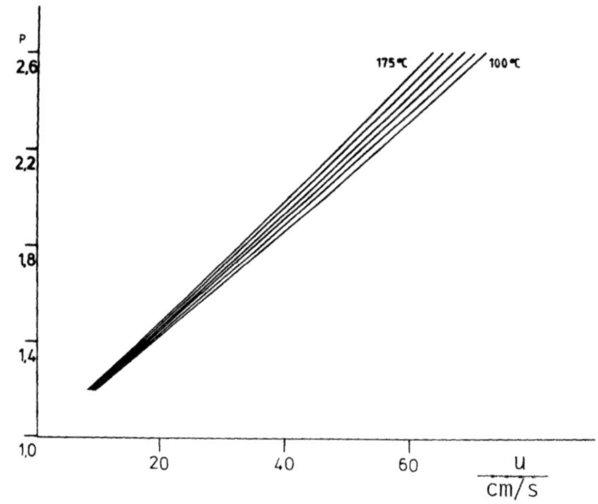

Fig. 8. Relation between adjusted pressure and average linear gas velocity at different temperatures.

Because of the linearity between average linear gas velocity, capillary column diameter, capillary column length, pressure and temperature, there is a possibility of obtaining each of the shown parameters by simple calculation. For practical work especially the calculation of the average linear gas velocity at certain temperatures is useful because the average linear velocity depends on the gas viscosity and decreases by increasing temperature. By injecting CH_4 at various temperatures the resulting average linear gas velocity can be calculated from the dead time measured. Measuring of gas flows with bubble gas flow meters is normally just possible at ambient temperature. However, if one measures the gas flow at ambient temperature and increases the column temperature, the resulting gas flow differs completely to that being measured at ambient temperature.

As shown in Fig. 9 there is also a possibility to calculate the absolute gas velocity at the capillary column end.

As shown in Table I it is difficult to choose the right capillary column wall material.

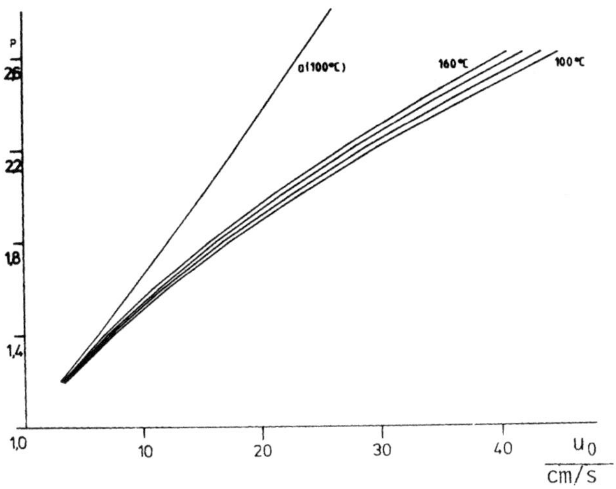

Fig. 9. Relation between gas velocity at the end of the column and pressure at different temperatures. Fused silica capillary WG-11 column (47 m x 0.22 mm i.d.) carrier gas, Nitrogen.

Table I. Influence of the capillary column wall material on the Index. a = deactivated by Carbowax 20M; b = deactivated by persilylation; c = deactivated by silylation D4; d = deactivated by polysiloxane degradation.

	WG-11					OV-101					
	X	Y	U	Z	S	X	Y	U	Z	S	
Soft glass	961	929	914	1159	1207	660	491	579	531	731	a
Pyrex glass	971	923	917	1167	1207	674	427	580	534	741	b
Duran glass	972	931	920	1170	1211	659	438	574	543	727	c
Fused silica	986	943	912	1169	1206	561	479	580	542	731	d

The Index is also influenced by the pretreatment of the capillary column wall material. For the choice of the right column wall material it should be kept in mind that:

- below 250°C there is no difference in column efficiency and inactivity between the three most used column wall materials: soft glass, borosilicate glass and fused silica. Polar phases should be used on soft glass, not at least for the price.
- polysiloxanes as stationary phases should be used generally on fused

silica up to 350°C;

- three-dimensional crosslinked stationary phases should not be used; there is hardly a necessity for these types of stationary phases, which are chemically bonded with the column wall material. They should only be used for fused silica and borosilicate glass;
- if really high temperature stable stationary phases are needed, it is necessary to use borosilicate glass.

From Fig. 10 the total range of the fragmentation of the polysiloxanes can be seen.

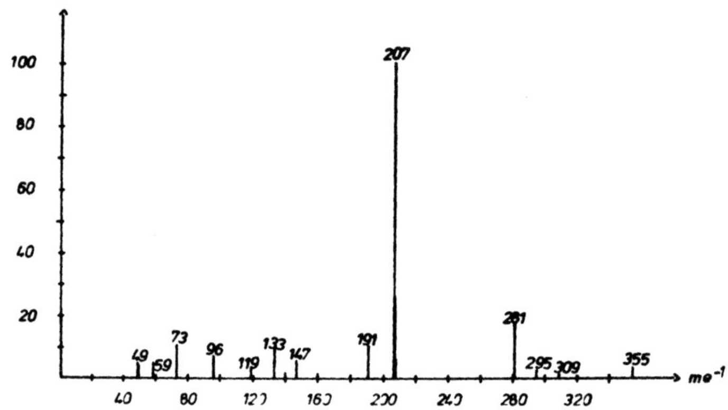

Fig. 10. Fragmentation of OV-1, three-dimensional crosslinked on Pyrex glass at 375°C. Ordinate: relative intensity (%).

The need for a chemical bonding with the glass surface is easy to recognize from Fig. 11, in which only two fragments can be seen.

Summarizing it can be stated that there is a need for all of the three column wall materials used today: soft glass, borosilicate glass and fused silica.

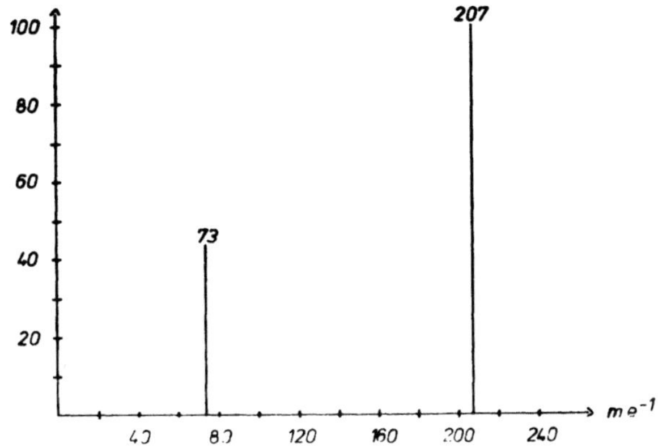

Fig. 11. Fragmentation of OV-1, three-dimensional crosslinked and chemically bonded to the Pyrex glass surface at 450°C. Ordinate: relative intensity (%).

GAS CHROMATOGRAPHIC CHARACTERIZATION OF FREQUENTLY OCCURRING SESQUITERPENES IN ESSENTIAL OILS

É. LEMBERKOVICS, G. VERZÁR-PETRI

ABSTRACT

Relative retention factors referred to guaiazulene as standard, were determined on two stationary phases of different polarity for the GC characterization of frequently occurring sesquiterpenes in essential oils. The standard deviation, calculated from the relative retention factors of the same sesquiterpene, determined in different oils, did not exceed 6×10^{-3}.

INTRODUCTION

Several sesquiterpenes (e.g. caryophyllene, cadinene, farnesene, bisabolene) are present in many essential oils (1). The aim of the present study was to determine the relative retention factors of these sesquiterpenes in order to facilitate their identification in essential oils.

EXPERIMENTAL

The experimental work was carried out on a gas chromatograph Jeol 1100 equipped with FID and glass columns. Column 1 (3 m x 2.3 mm i.d.) packed with 3% OV-17 on Gas-Chrom Q (100-120 mesh); other conditions: oven temperature, 60-230°C (8°C/min); carrier gas, Nitrogen, 38 ml/min; injector, 200°C; detector, 240°C. The efficiency referred to linalool was 1825 plates per m at Reynold's constants of 119, 158, 162, 243 and 202 (2). Column 2 (3 m x 3.4 mm i.d.) packed with 1.5% Sp-2250 + 0.95% Sp-2401 on Supelcoport (100-120 mesh); other conditions as for column 1. The efficiency referred to linalool was 1776 plates per m at Reynold's constants of 129, 189, 238, 330 and 244. The constants of the mixed phase were calculated from those of pure components (2).

The concentration of the investigated solution was 10-20 mg/ml of essential oil in chloroform or n-hexane. Samples of 0.1-0.5 µl were injected. The retention times were determined with an accuracy of 0.1 s; the relative retention factors were determined from 3-5 parallel measure-

A. Baerheim Svendsen and J.J.C. Scheffer (eds), Essential oils and aromatic plants, ISBN-13: 978-94-009-5137-2

ments; the accuracy of the oven temperature was ± 1°C. The sesquiterpenes in the essential oils were identified by MS, GC-MS or GC-IR. In a few cases (farnesene isomers, α-bisabolol) the compounds were isolated by preparative GC or by TLC prior to the MS or IR analysis (3-6).

The analyses were carried out with sesquiterpene-rich Hungarian commercial oils (chamomile, yarrow, basil and wormwood oil) and with the essential oil of Mexican *Cannabis sativa* grown in Hungary.

RESULTS AND DISCUSSION

Since no pure sesquiterpenes were available for the determination of their retention factors, the determination of those factors was performed on the basis of the data obtained from the sesquiterpenes occurring in the five essential oils mentioned above. The retention factors were determined on two stationary phases of different polarity (see Experimental), and guaiazulene was used as standard. In order to improve the reproducibility, also the retention factors of an adequately chosen n-alkane series (pentadecane - eicosane) were calibrated against guaiazulene. The standard deviation of the retention factors was determined for those sesquiterpenes which occurred in more than two essential oils; it did not exceed 6×10^{-3}.

The characteristic elution sequence of the sesquiterpenes was determined by means of the retention parameters on the stationary phases: in the range 160-180°C the sesquiterpene hydrocarbons, except the azulenes, were eluted, in the range 180-210°C the oxygen-containing sesquiterpenes and in the range 210-230°C the guaiane-type of sesquiterpenes. A better separation of some sesquiterpenes was obtained on the mixed stationary phase.

The following sesquiterpenes identified in our experiments, were not mentioned in previous publications (6-20): α-farnesene in Hungarian chamomile and cannabis oil, α-bisabolol in basil and wormwood oil, α-humulene, γ-cadinene, γ-muurolene and caryophyllenol I in yarrow oil.

The study of the retention parameters of the different oils on two stationary phases provided informations about the structure of non-identified components as well. Only when the relative retention factors of the sesquiterpene peaks on both stationary phases are the same, the compounds may be identical. If the retention factors differ on one of the phases, they are certainly not.

CONCLUSIONS

Relative retention factors of sesquiterpenes, determined in the same way as those of monoterpenes (2), facilitate the qualitative analysis of sesquiterpenes in essential oils.

ACKNOWLEDGEMENTS

We thank Dr.Sc. József Tamás, senior scientist at the Central Chemical Institute of Budapest and Mrs. I. Héthelyi, scientist at the Pharmacological Research Institute of Budakalász for their assistance with the MS analysis.

REFERENCES

(1) Gildemeister E, Hoffmann Fr. 1956-1961, Die ätherischen Öle, Band I-VI, Berlin, Akademie-Verlag.
(2) Lemberkovics É. 1984, J. Chromatogr. 286:293.
(3) Lemberkovics É. 1979, Sci. Pharm. 47:330.
(4) Lemberkovics É, Veszki P, Verzár-Petri G, Trka A. 1981, Sci. Pharm. 49:401.
(5) Lemberkovics É, Verzár-Petri G, Szabó G, Bihari M. 1982, Herba Hung. 21:197.
(6) Lemberkovics É, Thesis, Semmelweis Medical University, Budapest, Hungary, in press.
(7) Glasl H. 1972, Pharm. Ind. 34:122.
(8) Schilcher H. 1973, Planta Med. 23:132.
(9) Gasič O, Lukič V, Nikolič A. 1983, Fitoterapia 2:51.
(10) Bejnarowicz EA, Smolenskij SJ. 1968, J. Pharm. Sci. 57:2160.
(11) Shalaby A. 1978, Thesis, Semmelweis Medical University, Budapest, Hungary.
(12) Bercht CAL, Paris MR. 1973-74, Bull. Tech. Gattefosse 68:87.
(13) Kunde R. 1974, Thesis, University of Saarbrücken, FRG.
(14) Hendriks H, Malingré Th M, Batterman S, Bos R. 1975, Phytochemistry 14:814; 1978, Pharm. Weekbl. 113:413.
(15) Masada Y. 1976, Analysis of Essential Oils by Gas Chromatography and Mass Spectrometry, New York/London/Sydney/Toronto, J. Wiley & Sons.
(16) Bowers WS, Nishida R. 1980, Science 209:1030.
(17) Goraev MI, Bazalitskaya VS, Lishtvanova LN. 1962, Zhur. Prikl. Khim. 35:2799.
(18) Greger H. 1978, Phytochemistry 17:806.
(19) Kaul VK, Nigam SS, Banerjee AK. 1979, Indian Perfum. 23:1.
(20) Sayed MD, El-Shamy AM, Soliman FM, El-Shabrawy AO. 1979, Bull. Fac. Pharm., Cairo Univ. 16:85.

PROGRESS IN ISOLATION TECHNIQUES FOR ESSENTIAL OIL CONSTITUENTS

K.-H. KUBECZKA

ABSTRACT

The procedures most frequently applied to isolate the constituents of essential oils have been fractional distillation and chemical methods such as formation of decomposable derivatives. The major limitations of both techniques are that pure compounds are relatively seldom obtained and that isomerization and decomposition occasionally take place.

A substantial improvement was the introduction of chromatographic methods for essential oil analysis. Among these methods, which in the ideal case allow a separation of individual components, are first of all thin layer chromatography and preparative gas chromatography. Recently, also high performance liquid chromatography has been shown to be well suited for the separation of essential oil components and sometimes it shows significant advantages over TLC and GLC. Since the analysis generally is performed without exposition to air, as in the case of TLC, and at ambient temperature, the destructive temperatures needed in GLC are avoided and degradation products are not encountered.

In addition to HPLC the relatively newly developed DCCC and RLCC can be used for separating essential oils into fractions or in the ideal case into individual pure compounds. These methods, based on the partition of oil constituents in a biphasic system, allow above all the enrichment of minor constituents, since relatively large samples can be separated in one analytical run. Comparing RLCC and DCCC one can particularly stress the superior flexibility of RLCC because no droplet formation is needed and consequently solvent system in RLCC may be freely chosen.

All methods mentioned have their advantages and disadvantages and often only a skilful combination of various methods lead to the results wanted.

INTRODUCTION

Essential oils are usually chemically complex mixtures. Only few groups of naturally occurring products contain as many compounds as essential oils and a number of attempts have been made to optimize their analysis. By means of modern analytical techniques it has been possible to identify over 100 different components of an essential oil, mainly terpenes, phenylpropane derivatives and aliphatic compounds. Most oils contain one or a couple of main components, but also minor components may be important. Because of the large variations in the amounts of the various components, additional problems may arise in the separation. Relatively large amounts

A. Baerheim Svendsen and J.J.C. Scheffer (eds), Essential oils and aromatic plants, ISBN-13: 978-94-009-5137-2

of the sample have to be used in order to isolate trace components. Under such circumstances it is usually difficult, if not impossible, to isolate minor constituents by a single separation procedure, and often only a skilful combination of various separation techniques will lead to the anticipated results.

In this paper the isolation of essential oil constituents in the milligram-gram range is discussed, amounts which are sufficient for structure elucidation. For ^{1}H-NMR, ^{13}C-NMR, IR and MS analysis about 10-20 mg of a pure compound are needed, for the determination of the molecular formula by elementary analysis additionally 5 mg. In order to prepare chemical derivatives or to investigate the chemical reactivity of a compound, again 10-100 mg are necessary. It is usually still possible to separate such amounts by means of 'up-scaled' analytical methods. However, such a procedure may be both time-consuming and laborious.

The procedures which were mostly used during the last century for the isolation of essential constituents were fractional distillation and chemical methods. By formation of various decomposible and water soluble derivatives, dependent upon the functional groups, a preparative separation of an essential oil into several fractions was achieved (Fig. 1).

Fig. 1. Separation of an essential oil into different groups of components.

Essential oil
Aqueous sodium carbonate (5%) → acids
Aqueous sodium hydroxide (5%) → phenols
Methanolic sodium hydroxide (0.5 N) → esters, lactones
Sodium hydrogen sulphite or Girard reagent → aldehydes, ketones
Phthalic anhydride → prim./sec. alcohols
Residue → tert. alcohols, ethers, hydrocarbons, etc.

Acids, phenols as well as the relatively seldom occurring bases may be well separated by formation of water soluble salts. Esters and lactones can be hydrolyzed and the resulting acids separated as salts. Carbonyl compounds can be isolated after reaction with sodium hydrogen sulphite or Girard reagent, primary and secondary alcohols as phthalic acid esters.

The final separation of the remaining mixture, consisting of tertiary alcohols, ethers and hydrocarbons, as well as that of the fractions obtained, must be accomplished by physical methods, such as fractional distillation, or on the basis of different solubility. Otto Wallach recognized that natural terpene mixtures cannot be sufficiently separated by fractional distillation due to the similarity of their boiling points, and he introduced the formation of crystalline derivatives as separation and purification method.

FRACTIONAL DISTILLATION

Although it is only seldom possible to obtain chemically pure components from essential oils by fractional distillation, this method continues to play an important role in the preparative isolation of essential oil constituents. With the development of efficient column types, it became possible to fractionate milliliter amounts of essential oils under mild conditions. The various constituents of the fractions collected may subsequently be separated by using such a column.

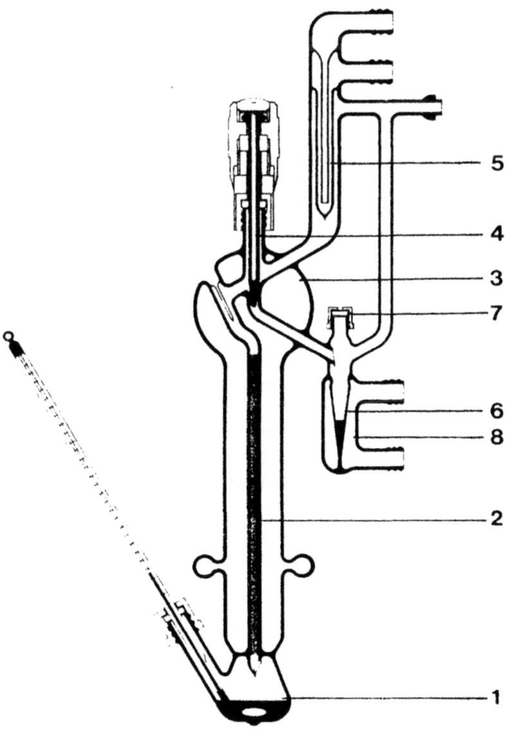

Fig. 2.

Cross-section of a 'Fischer Spaltrohr' column (reproduced with the permission of Fischer Labor- u. Verfahrenstechnik, Meckenheim, FRG):
1. distillation flask;
2. 'Spaltrohr' column;
3. still head;
4. needle valve;
5. reflux condenser;
6. distillate receiver;
7. sampling port for taking out samples under reduced pressure;
8. jacket for cooling or heating the distillate.

By means of a so-called 'Spaltrohr'-column (Fig. 2), which has its name from the small space between two concentric glass tubes, amounts of 0.1 ml upto a few milliliters may be distilled at a pressure below 1 mbar, thus allowing the distillation of thermally unstable components. With this column up to 100 theoretical plates may be achieved (1).

The separation of citronellyl acetate and geranyl acetate with boiling points of 73.5^{0}C and 77.0^{0}C, respectively, at approximately 1 mbar clearly demonstrates the efficiency of the device (Fig. 3).

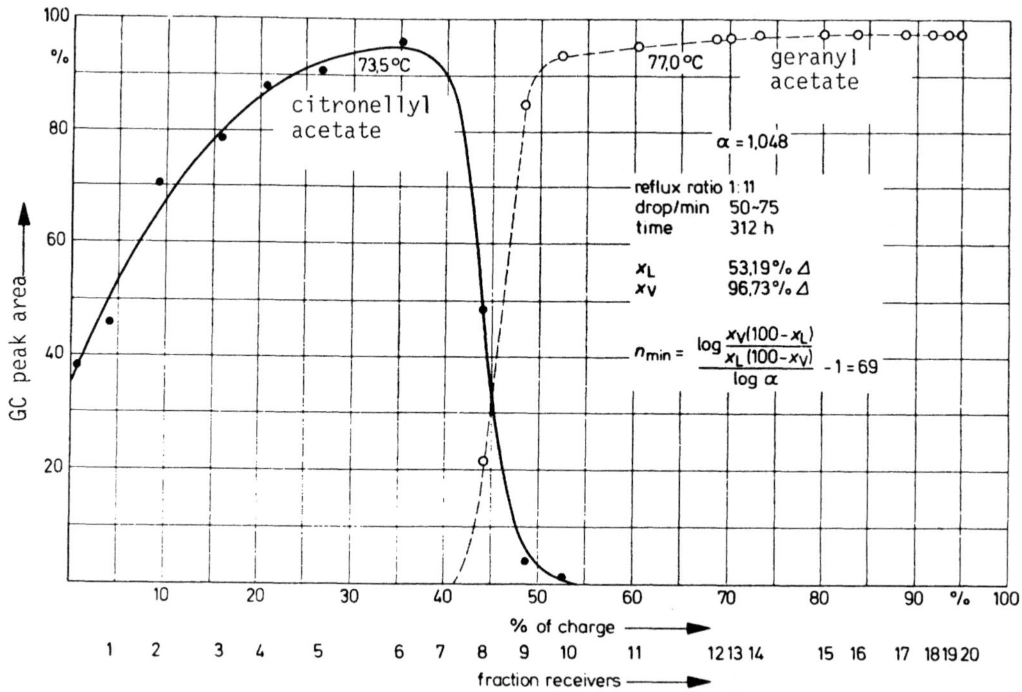

Fig. 3. Separation of citronellyl acetate and geranyl acetate by means of a 'Spaltrohr' distillation column (reproduced with the permission of Fischer Labor- u. Verfahrenstechnik, Meckenheim, FRG).

The major limitation of fractional distillation is that pure compounds relatively seldom are obtained from essential oils and that isomerization and decomposition of labile components occasionally take place.

LIQUID COLUMN CHROMATOGRAPHY

The most common method of fractionation of essential oils is the separation of their hydrocarbons and oxygen-containing components according to Kirchner and Miller (2). The hydrocarbons are separated on silica gel by elution with pentane or hexane. However, several authors have found rearrangements of some terpene hydrocarbons during the chromatography. Active sites on dried silica gel and impurities of metals have been found to be responsible for a series of isomerizations of a number of oil constituents. By using pure silica gel and deactivation of it by adding 5-7% of water (3, 4) isomerization processes could be avoided. After desorption of the non polar components from the column, the oxygen-containing constituents can be eluted in order of increasing polarity by applying a gradient elution with solvents such as diethyl ether, ethyl acetate, chlorinated hydrocarbons and methanol in pentane or hexane, yielding a number of fractions, and possibly some pure components. Similar results were obtained by other investigators using alumina, but silica gel is the most suitable sorbent for essential oils (5-7).

An efficient fractionation in terms of speed and simplicity has been developed (4), using silica gel dry-column chromatography. The procedure which has proved useful for prefractionation of an essential oil, ensures to a high degree the avoidence of artifact formation and allows a preliminary separation into five fractions of different polarity. Two slightly polar fractions are obtained by elution with pentane and benzene. Instead of benzene we prefer now the non toxic isopropyl chloride, which has similar polarity and lower boiling point.

The polar fractions of the mixture are obtained by cutting the column into three parts and eluting with diethyl ether-methanol 8:2 (v/v). By means of standardized silica gel for dry-column chromatography with an adjusted water content of 7%, an additional deactivation of the sorbent is not necessary. Fraction 1 contains the non polar compounds (hydrocarbons) and fraction 2 the slightly polar compounds, such as esters of monocarboxylic acids of medium chain-length. Compounds found in fraction 3 are among others acetates of terpenes and aliphatic alcohols. Fraction 4 contains aldehydes and ketones and fraction 5 the most polar compounds, such as alcohols.

A further separation of the fractions collected by LC can be achieved by applying other chromatographic methods. The non polar fraction, mainly consisting of monoterpene and sesquiterpene hydrocarbons, yields after fractional distillation uniform C10 and C15 fractions, which can be subjected to LC over silica gel impregnated with silver nitrate.

Instead of silica gel, a macroporous ion exchanger (Lewatit SP 1080, Merck) in Ag^+ form is recommended for the separation of unsaturated compounds (8). The advantage of such a technique seems to be that only π-complexes of the silver ions are formed with olefin bonds without further adsorption phenomena. Since the terpene hydrocarbons are efficiently separated according to the character and number of the double bonds, specific separations may be obtained. Sometimes, however, isomerizations of proton sensitive terpenes have been observed, e.g. isomerization of cubebene to a diene hydrocarbon (9) caused by the acidic reaction of decomposed silver nitrate.

We have, therefore, investigated alternative procedures and have found dry-column chromatography on silica gel with n-pentane well suited for the

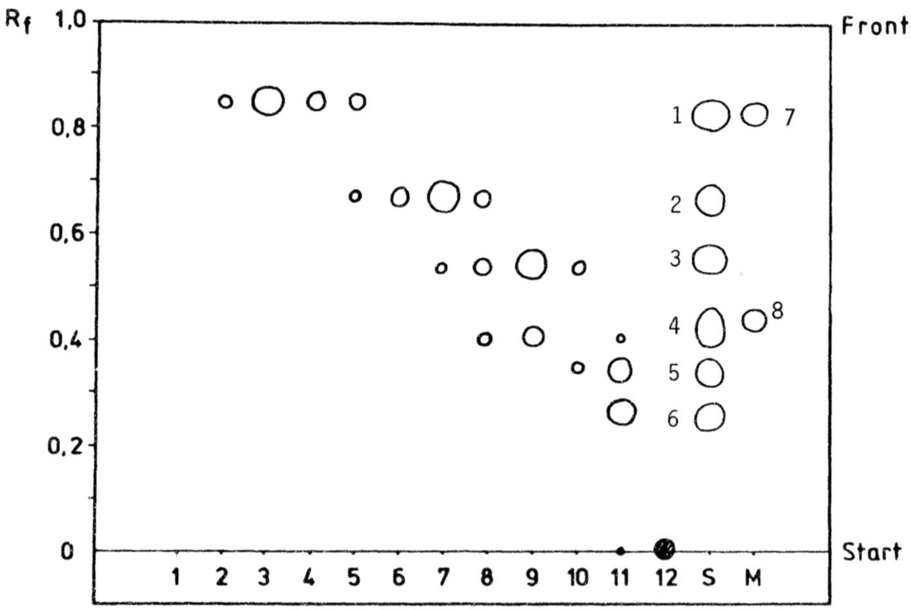

Fig. 4. TLC (silica gel/n-pentane at -18°C) of the 12 fractions obtained by silica gel dry-column chromatography with n-pentane at -18°C. S and M = sesqui- and monoterpene hydrocarbons: 1 = α-copaene; 2 = δ-elemene; 3 = β-elemene; 4 = β-caryophyllene; 5 = β-bisabolene; 6 = α-humulene; 7 = α-pinene; 8 = limonene.

separation of terpene hydrocarbons, if it was carried out at low temperatures (10). Using a 50 cm x 40 mm column packed with 225 g silica gel, corresponding to 36 cm height, a fairly good separation of 6 ml terpene hydrocarbons could be achieved with 320 ml n-pentane at -18°C (11). Fig. 4 shows a low temperature TLC analysis of the various fractions obtained after cutting the column into 12 parts, each of 3 cm, and eluting with n-pentane after addition of some water.

Partition chromatography has also been applied in a few cases for the separation of terpenes or terpene fractions, but due to the unstable coating of the support with stationary liquids, the method has not been used often. The use of chemically bonded phases, known as 'reversed phases' in partition chromatography, is becoming increasingly important in the separation of essential oil constituents by HPLC, as will be discussed later.

An example of a separation of sesquiterpene alcohols from the essential oil of an *Artemisia* species by using different types of columns in series, is given by Jork and Juell (12).

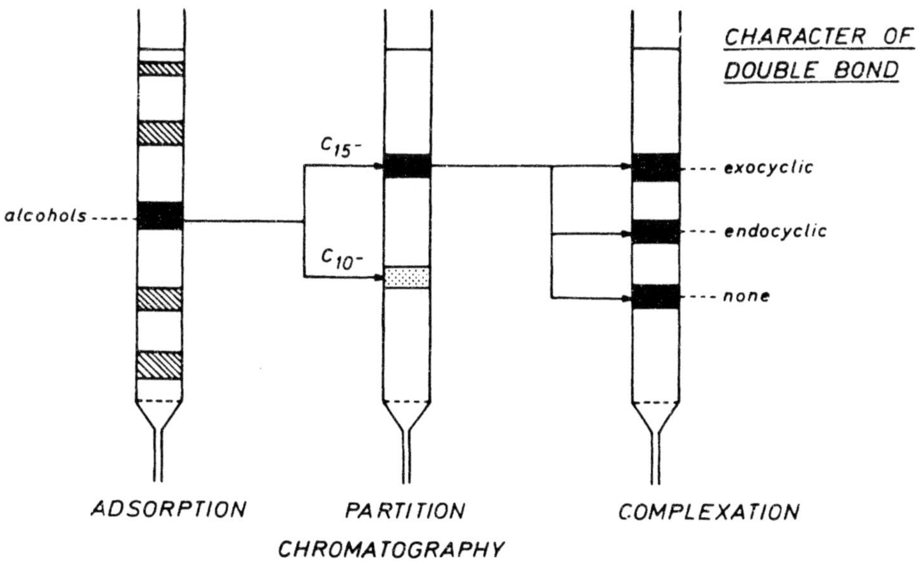

Fig. 5. Separation of sesquiterpene alcohols from an *Artemisia* oil by using different types of columns in series (12).

After adsorption chromatography of the total essential oil (Fig. 5) the alcohol fraction was subjected to partition chromatography. Finally the sesquiterpene alcohol fraction was separated into individual compounds by complexation chromatography, using silver nitrate impregnated silica gel.

THIN-LAYER CHROMATOGRAPHY

Due to its simplicity and speed TLC is along with GC still one of the most important methods in the analysis of essential oils, as can be seen from a number of papers (13, 14). Because of the speed of this fairly

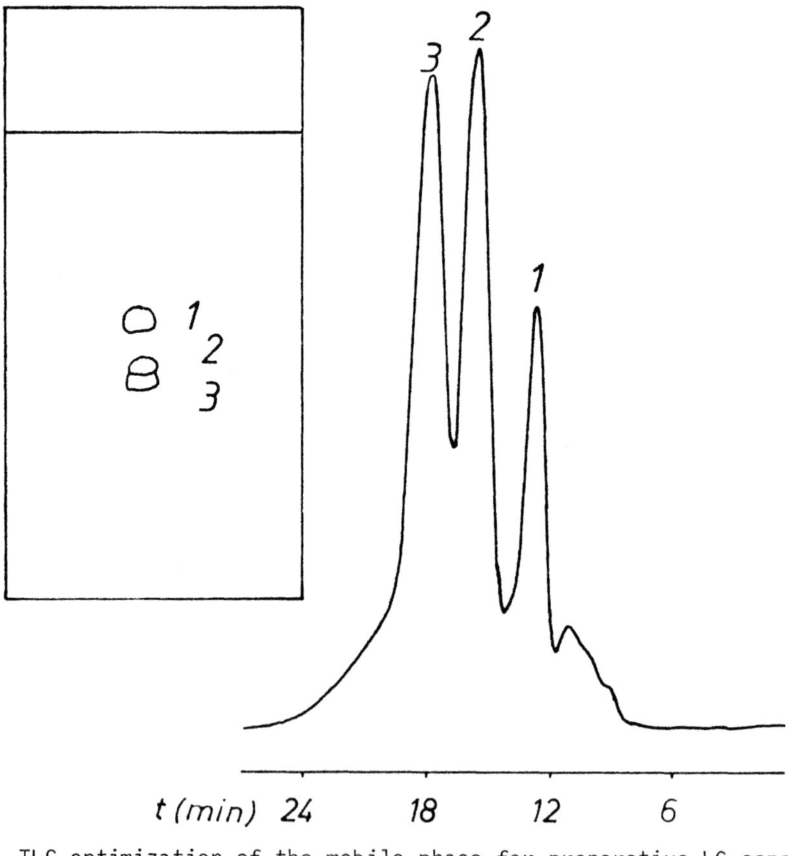

Fig. 6. TLC optimization of the mobile phase for preparative LC separation of *Dictamnus albus* fruit oil constituents. 1 = methyl cinnamate, 2 = neral, 3 = geranial.

cheap method, a suitable solvent system may be found for column chromatographic separations by a few TLC runs. Therefore, TLC is often used as a pilot technique for LC separations of essential oils (15, 16). Fig. 6 shows a thin-layer chromatographic optimization of the mobile phase for a preparative LC separation of oxygen-containing terpenes from the fruit oil of *Dictamnus albus*. In order to obtain a good LC separation, the mobile phase has to be adjusted by TLC to give R_f-values between 0.35 and 0.65.

In a slightly modified form TLC can also be used for preparative separations. By means of plates up to 1 m long and a layer thickness up to 2 mm, several grams of a mixture of compounds can be applied as a line. To obtain good resolution of narrow bands, repetitive development is recommended (17, 18). Because of the sharper separations, preparative TLC is often preferred to conventional LC. The various bands are scraped off, collected and the components eluted with appropriate solvents. Since the surface of the layer is normally exposed to air during the evaporation of the solvent, volatile components such as monoterpene hydrocarbons can be lost at a significant rate, and oxygen sensitive compounds can be decomposed.

GAS CHROMATOGRAPHY

Along with LC, GC has become one of the most important tools in the analysis of essential oils and it plays today a central role in the study of volatiles. In GC the sample must be gasified, but volatility is just a characteristic for essential oils. It is therefore not surprising that only a few years after the introduction of GC, experiments on preparative separations of essential oils were conducted. Since with an analytical packed column only small amounts can be analyzed, the most obvious way to increase the capacity of a column is to increase the diameter. This allows the sample size to be increased as the square of the column radius. Another possibility is to operate under overloaded conditions, but this requires well resolved peaks and makes it difficult to separate mixtures of components with close retention times. Therefore, repetitive injections of smaller samples under optimum conditions, using efficient columns with internal diameters of 6-8 mm prove to be the best solution. But of three desirable properties only two can be realized at the expence of the third one when using long, small diameter columns, as pointed out by Dimick (19):

'Good resolution and reasonable speed call for small samples. Large samples, well resolved, require long separation times. Rapid separation of large sample excludes high resolution'.

A serious problem in preparative GC of essential oils is encountered because of the formation of aerosols. Different types of collector bottles or glass tubings, containing glass wool have, therefore, been described for the collection of aerosols. Aerosol formation can, however, be avoided right from the beginning, if an appropriate cooling temperature is chosen. The best results were obtained by applying a flat temperature gradient, as proposed by Armitage (20).

HIGH PERFORMANCE LIQUID CHROMATOGRAPHY

HPLC has recently been shown to be well suited for preparative separation of essential oils (21). Sometimes it shows significant advantages over TLC and GC. Since the analysis is generally performed without exposition to air, as in TLC, and at ambient temperature and not at relatively high temperatures, as in GC, degradation products are not encountered.

The use of HPLC for the separation of essential oils has already been tried sporadically, but probably the good separation obtained by GC in this field has delayed the application of HPLC. Another important restricting factor is the commercially available HPLC detectors. Many essential oil components cannot be analyzed by HPLC with UV detection at 254 nm because of lack of chromophoric groups. Therefore, HPLC was not considered useful for the analysis of essential oils, but rather for quality control of mixtures of compounds, having good chromophoric groups. In such cases HPLC gives results quite quickly and, because of the selectivity of the UV detector, a much simpler chromatogram is obtained than with GC. In addition, HPLC enables in contrast to GC, separation of non volatile compounds. An important advantage of HPLC is also the sample recovery. The advantages mentioned prompted us to investigate the applicability of HPLC for preparative separation of essential oils, especially the mono- and sesquiterpenes.

Because of lack of chromophoric groups in most of these compounds, detection by means of refractive index (RI) or low UV monitoring is necessary. Since RI detection goes to the account of sensitivity, we preferred low UV detection. As demonstrated earlier, all constituents of a mixture of natural, volatile terpenes can be detected by their UV absorbance from

200 to 220 nm (23, 24). This limits, however, the application of several solvents and it requires mobile phases which do not exhibit too high background absorption at these wavelengths.

Another problem in HPLC analysis of essential oils was the restricted capacity and the relatively small range of k'-values of a LC system. Therefore, an effective separataion of multicomponent mixtures in one operation was difficult. A preliminary LC fractionation was, therefore, necessary. Several less complex fractions were collected, which subsequently could be separated into their components.

The separation of complex mixtures of different classes of compounds, such as hydrocarbons, esters, ketones, aldehydes and alcohols, is usually achieved by LC using silica gel and solvents of increasing polarity. The technique fails, however, in the separation of the mono- and sesquiterpene hydrocarbons in separate fractions. Thus, in that case, an additional separation is usually necessary. In the investigations described here (24) we found that a mixture of oxygen-containing terpenes, as well as mono- and sesquiterpene hydrocarbons could be fractionated satisfactorily in 12 min on an octadecyl-

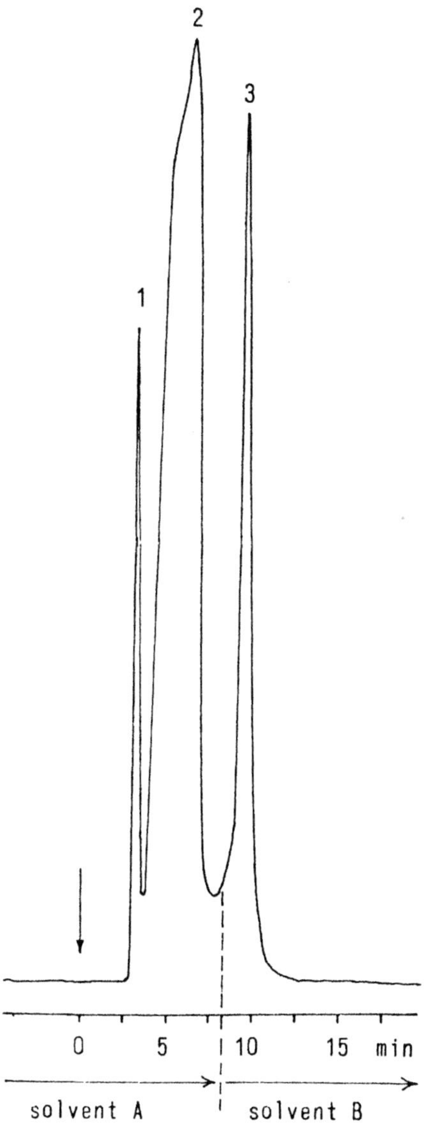

Fig. 7. HPLC fractionation of an essential oil at a flow rate of 8 ml/min.
Conditions: column, 24 cm x 10 mm i.d., LiChroprep RP 18 (40 µm); mobile phases, A) methanol-water 82.5:17.5 (v/v), B, methanol; flow rate, 8 ml/min; detector, UV 220 nm.
1 = oxygen-containing compounds,
2 = monoterpene hydrocarbons,
3 = sesquiterpene hydrocarbons.

silane-bonded silica phase (LiChroprep RP 18, 40 µm particles). The HPLC system was operated on a semi-preparative scale with a resolution which was much better than the prefractionation on an ordinary silica gel column. With a mobile phase of methanol and water, which allows low UV monitoring, a good separation of the three groups was achieved. By stepwise elution with methanol-water 82.5:17.5 (v/v) and pure methanol it was possible to separate up to 0.5 ml of an essential oil. The 24 cm x 10 mm i.d. column was operated at an eluent flow rate of 8 ml/min. The elution order of the investigated compounds took place according to a decreasing polarity, and within the group of hydrocarbons according to increasing molecular weight (Fig. 7). The elution order was confirmed by capillary GC of each fraction: Fraction 1 contained indeed the oxygen-containing mono- and sesquiterpenes, fraction 2 the monoterpene hydrocarbons and fraction 3 the sesquiterpene hydrocarbons, even labile compounds such as germacrene-D.

At lower flow rates with the RP 18 HPLC column, a separation within the three fractions, especially within the oxygen-containing compounds, was observed. This goes, however, to the account of the separation time. Reduction of the flow rate of the mobile phase from 8 ml/min to 4 ml/min increased the elution time from 12 min to about 30 min, but led to a resolution of the first HPLC fraction, containing the oxygen-containing terpenes into four peaks (Fig. 8). GC indicated that the monoterpene alcohols were eluted in peak 2, that peak 3 contained the less polar monoterpene alcohol esters, peak 4 the sesquiterpene alcohols and peak 1 only solvent impurities.

Summarizing, it may be concluded that the described semi-preparative HPLC separation into different groups was surprisingly good, considering the complexity of the natural terpene mixture investigated. HPLC is, thus, well suited for a fast fractionation, required for an efficient separation of such complex mixtures containing chemically different groups of constituents.

Further separation of the hydrocarbon fractions was performed by means of LSC, since experiments with different reversed phase materials had failed in a satisfactory separation of the monoterpenes. We used n-pentane as mobile phase and irregular 7 µm particles of silica gel (LiChrosorb Si 60) as adsorbent.

The water content of the adsorbent influences strongly the separation. Dry or fully activated adsorbents can cause sample alteration. Therefore, it was important to add a certain amount of water to the adsorbent before

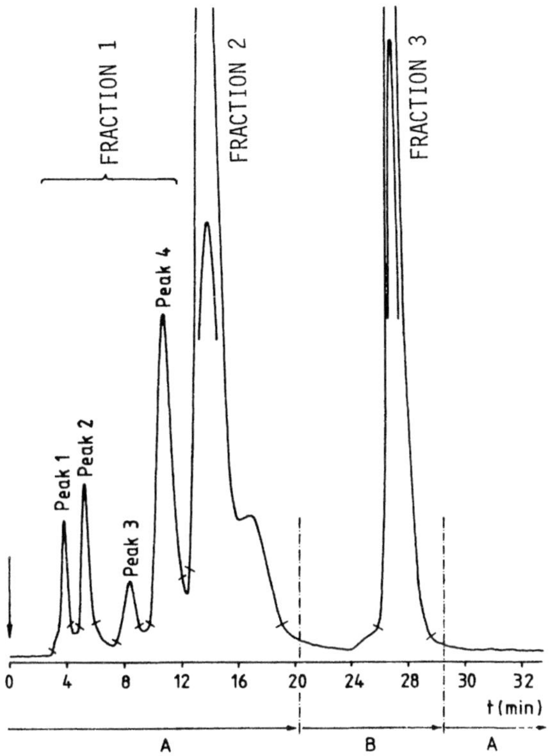

Fig. 8. HPLC fractionation of an essential oil at a flow rate of 4 ml/min. Other conditions, see Fig. 7.

use. Since the k'-values decrease when more water is added, the optimum percentage has to be exactly evaluated; it was found to be 4.8%. Sample alteration and irreversible adsorption, which were observed for terpenes in classical LSC due to the active surface of the adsorbent, do not occur on silica gel with a water content of more than 4% (3).

Another factor which controls the k'-values of the different hydrocarbons, is the temperature, as it is known from TLC. The k'-values increase with decreasing temperature. Fig. 9 shows the resolution r versus the column temperature T for monoterpene hydrocarbons. It can be seen from the figure that the separation and the plate number n improve with decreasing column temperature. This is in agreement with the observations of Jinno et al. (25) and of Snyder (26). So, terpene hydrocarbons can be

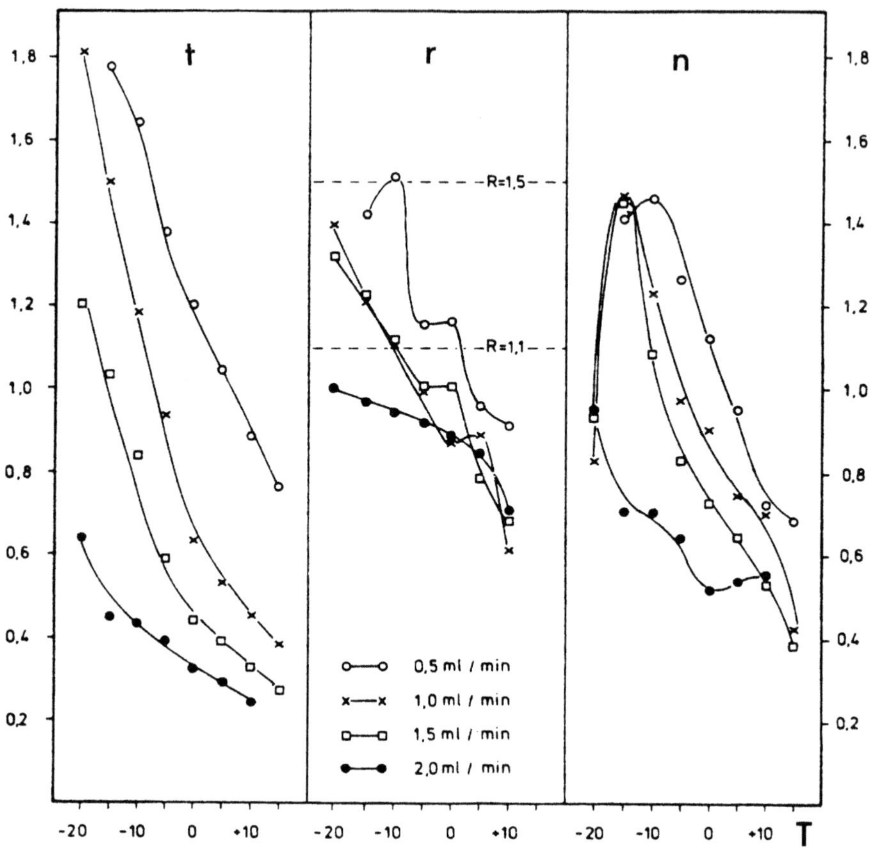

Fig. 9. Analysis time t, resolution r and plate number n versus column temperature T in adsorption HPLC of terpene hydrocarbons on silica gel with n-pentane.

sufficiently separated only at low temperature. We preferred -15°C (10), since a lower temperature resulted in too long and, therefore, ineffective separations without significantly better resolution.

The separation of eight monoterpene hydrocarbons achieved in about 10 min under the conditions described above, is shown in Fig. 10. Under the same chromatographic conditions a natural mixture of sesquiterpene hydrocarbons was separated in less than 20 min (Fig. 11).

The combination of a reversed-phase HPLC fractionation followed by a further HPLC separation, provided in many cases pure compounds, the structure of which could be determined with various spectroscopic methods.

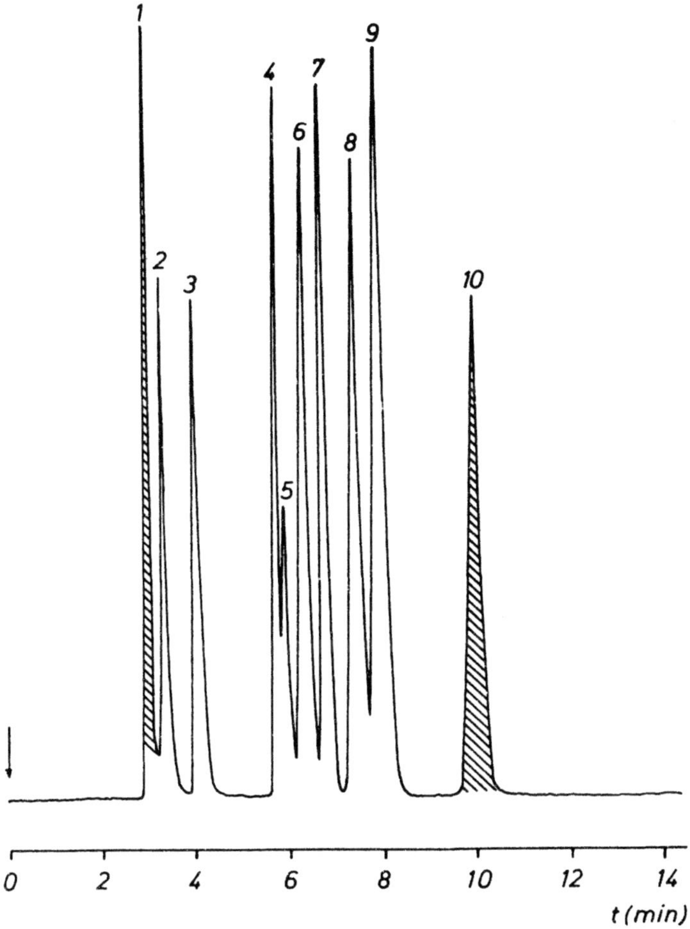

Fig. 10. HPLC separation of monoterpene hydrocarbons. Conditions: column, 30 cm x 4 mm i.d.; adsorbent, LiChroprep Si 60 (7 µm) with 4.8% water; eluent, n-pentane, 1.5 ml/min; temperature -15°C; UV detector. 1 = standard, 2 = α-pinene, 3 = β-pinene, 4 = sabinene, 5 = limonene, 6 = Δ3-carene, 7 = myrcene + cis-β-ocimene, 8 = trans-β-ocimene, 9 = p-cymene, 10 = standard.

In some cases, however, further separation was necessary; it was usually conducted by subsequent LC or preparative GC.

A preparative HPLC separation of an essential oil, using a large diameter column, packed with octadecylsilan-bonded silica has recently been demonstrated (27). The essential oil of *Curcuma xanthorrhiza* was separated under isocratic conditions with methanol as mobile phase. On a column with

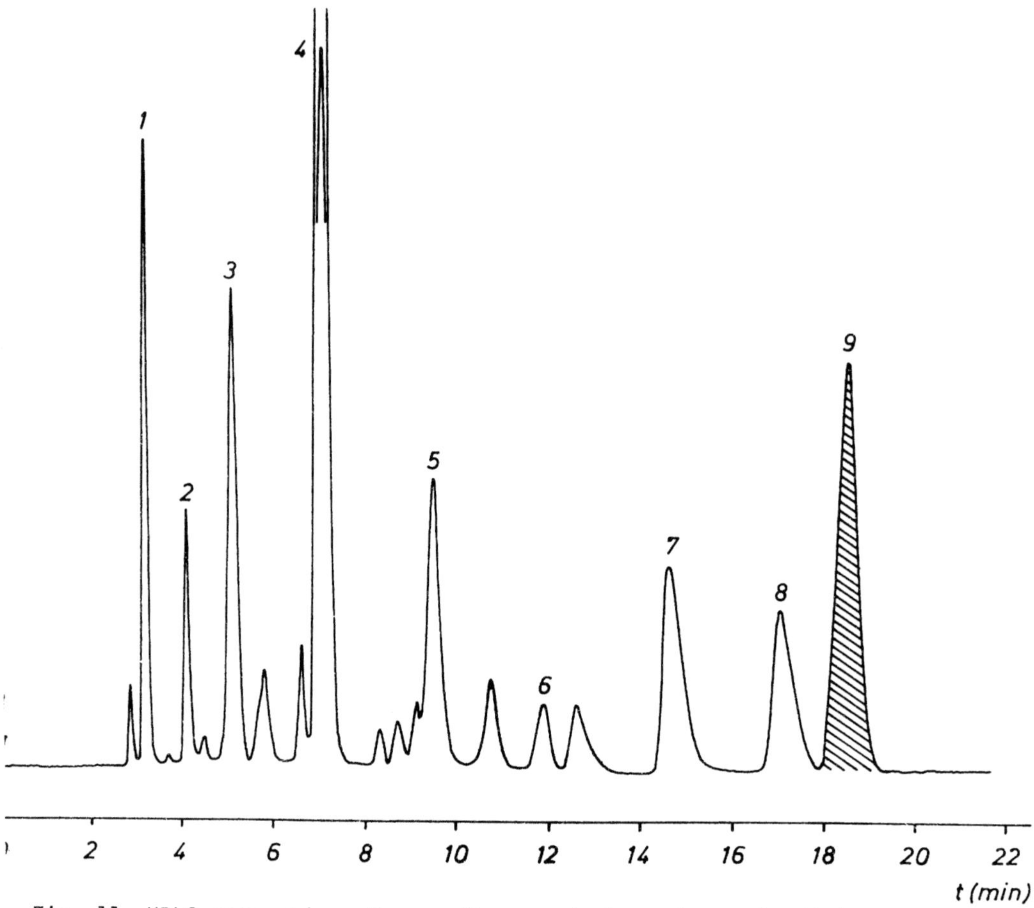

Fig. 11. HPLC separation of sesquiterpene hydrocarbons. Conditions see Fig. 10. 1 = α-copaene, 2 = δ-elemene, 3 = β-elemene, 4 = β-caryophyllene, 5 = α-bergamotene, 6 = β-bisabolene, 7 = α-humulene, 8 = δ-cadinene, 9 = standard.

an i.d. of 50 mm, 5 ml of the essential oil could be analyzed in one run and at least seven well separated fractions were collected within 12 min.

DROPLET COUNTER-CURRENT CHROMATOGRAPHY

In addition to HPLC the relatively newly developed technique of DCCC can be used for separating essential oils into fractions or, in the ideal case, into pure compounds. The apparatus which was developed by Tanimura et al. (28) and widely applied by Hostettmann (29), is suitable especially for the separation of relatively polar compounds.

Table I. DCCC fractionation of a terpene mixture. Percentages (x 10) of components in the fractions (underlined values indicate maximum amount).

Compound	Fractions																% in
	8-15	16-39	40-51	52-61	62-68	69-74	75-79	80-82	83-89	90-96	97-104	105-111	112-116	117-121	122-133	134-140	mixture
Limonene	86	627															9.29
1.8-Cineole			980	83													9.22
cis-β-Ocimene		28															0.19
trans-β-Ocimene		74															0.43
p-Cymene		13															0.07
Terpinolene	3	21															0.08
Linalool														445	641	204	10.68
Linalyl acetate					322	784	911	855	598	14	5	9					21.75
β-Caryophyllene	655	3															3.72
Humulene	77	6															0.45
α-Terpineol														16	120	224	2.24
Geranyl formate												53	208	25			0.24
Neryl acetate							1	6	94	236	115						0.87
Geranyl acetate						1	1	7	168	574	385	28	63				2.16
Caryophyllene oxide					89	44	19	8									0.65
Farnesol										29	337	566	264				1.61
Bisabolol					258	60	12	3									1.14
Elemol					176	53	14	3									0.88
Unknown u1													81	78	2		0.09
Unknown u2													126	59	1		0.05
Unknown u3													26	100	12		0.18
Unknown u4													133	150	5		0.18
Unknown u5													55	15			0.01

The apparatus consists of 300-600 columns filled with a stationary liquid. Separation is achieved by passing droplets of the mobile phase through the columns, thus distributing the components of the mixture at different ratios, which leads to their separation. With a water-free solvent system the separation of non-polar mixtures such as essential oils could be achieved (30-32). The fractionation of a terpene mixture by DCCC and checked by capillary GC is given in Table I.

Along with the separation of essential oil components, the method allows also an enrichment of minor components, since relatively large samples can be separated in one analytical run.

ROTATION LOCULAR COUNTER-CURRENT CHROMATOGRAPHY

Recent experiments with RLCC demonstrated clearly its usefulness for the separation of essential oils on semi-preparative scale (33).

The RLCC apparatus (Rikakikai Co., Tokyo, Japan) illustrated schematically in Fig. 12, consists of 16 concentrically arranged glass tubes, which are connected to each other in series with teflon tubing. The glass tubes are divided by teflon discs with a small hole in the center, thus creating small compartments or locules. After filling the tubes with the stationary liquid phase and completely expelling all air bubbles, the tubes are inclined to a 30^{0} angle.

In the ascending mode, the lighter mobile phase is applied to the bottom of the first tube (Fig. 12), thus displacing the stationary phase as its volume attains the level of the hole in the disc. The mobile phase passes through this hole and enters to the next compartment, where the process continues, etc., until the mobile phase emerges from the uppermost locule. Finally the two phases fill approximately half of each compartment. The dissolved essential oil subsequently introduced is subjected to a multistep partitioning process which leads to separations of the various components.

If the descending mode is selected for separation, the valve (Fig. 12) has to be switched, thus applying the (heavier) mobile phase at the top of each compartment.

When comparing RLCC and DCCC, one can particularly stress the superior flexibility of RLCC. Whereas DCCC under all circumstances requires a two-phase system able to form droplets of the mobile phase in the stationary phase, the choice of solvent-systems in RLCC is nearly unlimited. So, the limitations of DCCC, when analyzing lipophilic samples, do not apply to RLCC. Another advantage of RLCC is that one easily can change the stationary phase to become the mobile phase and vice versa. By doing so, the separation time can be drastically shortened without loss of separation efficiency. Even strong differences in polarity can easily be controlled by this change-over; this is not possible in DCCC.

The separation of groups of components with different polarities, even a separation of such similar components as menthone and iso-menthone, can be achieved.

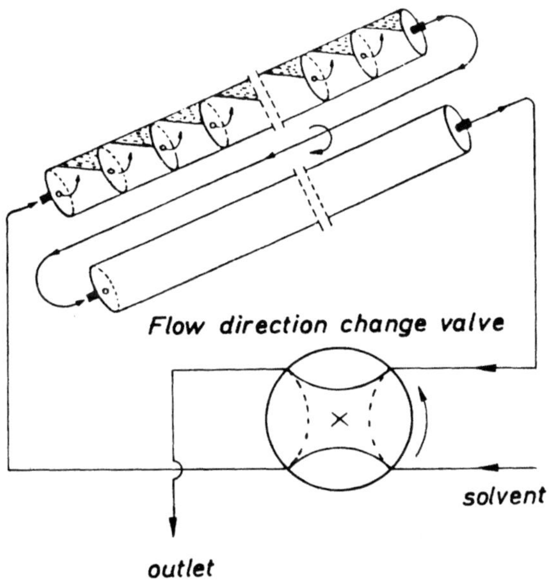

Fig. 12. Flow scheme of the RLCC apparatus.

Summarizing, an overview is given on different isolation techniques used in the field of essential oil analysis. All methods mentioned have their advantages and disadvantages, and only a skilful combination of various techniques, based on different separation characteristics, can usually lead to the results wanted.

REFERENCES

(1) Fischer WG. 1979, Labor Praxis 3:30.
(2) Kircher JG, Miller JM. 1952, Ind. Eng. Chem. 44:318.
(3) Scheffer JJC, Koedam A, Baerheim Svendsen A. 1976, Chromatographia 9:425.
(4) Kubeczka K.-H. 1973, Chromatographia 6:106.
(5) Motl O. 1975, in: Liquid Column Chromatography (Deyl Z, Macek K, Janák J, Eds.), Amsterdam, Elsevier Scientific Publ. Comp., p. 623.
(6) Scheffer JJC, Koedam A, Schüsler MThIW, Baerheim Svendsen A. 1977, Chromatographia 10:669.
(7) Mikes O. 1979, Laboratory Handbook of Chromatographic and Allied Methods. Chichester, Ellis Horwood Ltd.
(8) Houx NWH, Voerman S, Jongen WMF. 1974. J. Chromatogr. 96:25.
(9) Vlahov R, Holub M. 1967 (cit. from Heftmann E, Chromatography, New York, Reinhold Publ. Corp., p. 527).
(10) Kubeczka K-H, Schwanbeck J. 1977, Planta med. 32A:39.
(11) Schwanbeck J. 1981, Einsatzmöglichkeiten der HPLC bei der Analyse ausgewählter Apiaceenöle. Thesis, University of Würzburg, FRG.
(12) Jork H, Juell SM-K. 1979, Arch. Pharm. 312:540.
(13) Jänchen D. 1965-1984. Thin Layer Chromatography, Muttenz, CBS Camag Bibliography Service.
(14) Stahl E. (Ed.) 1967, Dünnschicht-Chromatographie. Ein Laboratoriumshandbuch. Berlin/Heidelberg/New York, Springer-Verlag.
(15) Jork H, Reh E, Wimmer H. 1981, GIT Fachz. Labor. 25:566.
(16) Koch V, Kubeczka K-H. 1984, Fresenius' Z. Anal. Chem. 318:243.
(17) Halpaap H. 1963, Chem.-Ing.-Tech. 35:488.
(18) Halpaap H. 1965, Chem. Ztg. 89:835.
(19) Dimick KP. 1966, G.C. Preparative Separations. Palo Alto, California, Varian Aerograph.
(20) Armitage F. 1969, J. Chromatogr. Sci. 7:190.
(21) Kubeczka K-H. 1981, in: Flavour'81 (Schreier P, Ed.), Berlin/New York, Walter de Gruyter, p. 345.
(22) Schwanbeck J, Kubeczka K-H. 1979, in: Königsteiner-Chromatographie-Tage 1979 (Kaiser UJ, Franzen KH, Eds.), Königstein, Waters GmbH.
(23) Strack D, Proksch P, Gülz P-G. 1980, Z. Naturforsch. 35c:675.
(24) Schwanbeck J, Koch V, Kubeczka K-H. 1982, in: Ätherische Öle, Analytik, Physiologie, Zusammensetzung (Kubeczka K-H, Ed.), Stuttgart/New York, G. Thieme Verlag, p. 70.
(25) Jinno K, Nomura H, Hirata Y. 1980, J. High Resol. Chromatogr. & Chromatogr. Commun. 3:305.
(26) Snyder LR. 1979, J. Chromatogr. 178:167.
(27) Matusch R, Pietschmann P. 1983. Schnelle Isolierung von Naturstoffen in pharmakologisch interessanten Mengen am Beispiel von *Curcuma* (DV). Lecture, Symposium 'Fortschritte in der Arzneimittelforschung', Munich.
(28) Tanimura T, Pisano JJ, Ito Y, Bowman RL. 1970, Science 169:54.
(29) Hostettmann K. 1980, Planta med. 39:1.
(30) Becker H, Hsieh WC, Verelis CO. 1981, GIT Fachz. Labor., Suppl. Chromatographie 81, p. 38.
(31) Becker H, Reichling J, Hsieh WC. 1982, J. Chromatogr. 237:307.
(32) Becker H. 1984, Fresenius' Z. Anal. Chem. 318:225.
(33) Kubeczka K-H. 1983, Isolierung ätherischer Ölbestandteile mittels RLCC. Lecture, 14th Int. Symposium on Essential Oils, Freising-Weihenstephan, FRG.

THE INFLUENCE OF ISOLATION CONDITIONS ON THE COMPOSITION OF ESSENTIAL OILS CONTAINING LINALOOL AND LINALYL ACETATE

G. SCHMAUS, K.-H. KUBECZKA

ABSTRACT

The essential oils of lavender and clary sage isolated by different techniques such as hydrodistillation or extraction with an organic solvent mixture, show significant differences. Systematic investigations on the stability of the major oil components, linalool and linalyl acetate, in a model system with buffered distillation water of different pH values enabled us to correlate the artifacts found in the hydrodistillate with linalyl acetate. Whereas linalool did not decompose in the pH range 5.0-8.0, an allylic rearrangement of linalyl acetate could not be prevented even at pH 7.0.

Since it was not possible to obtain by hydrodistillation an essential oil that represented the genuine composition, the volatiles of lavender and clary sage were primarily extracted with an organic solvent mixture. The non-volatile components of the extracts (fats, waxy materials and plant pigments) may render the GC analyses difficult. We applied therefore a gas phase stripping for the isolation of the volatiles. The combination of solvent extraction and gas phase stripping yielded an essential oil, the composition of which was qualitatively and quantitatively comparable to the genuine one, since artifact formation was prevented.

INTRODUCTION

GC is the most commonly used analytical technique in qualitative and quantitative studies on essential oils. In many investigations the oil sample was obtained by hydrodistillation, since sufficient amounts of oil can easily be isolated in that way. However, hydrodistillation can cause artifact formation of labile constituents, due to the temperature applied as well as to organic acids and metal ions in the distillation water. It should be pointed out that an artifact formation during the isolation can never be corrected at any later stage of the analysis.

It is well known that essential oil components such as linalool and linalyl acetate are labile and may undergo rearrangements during the isolation (1, 2). Therefore, we used fresh plant material of lavender and clary sage for a comparison of isolation procedures, since the compounds mentioned are major constituents of the essential oils of lavender and clary sage.

A. Baerheim Svendsen and J.J.C. Scheffer (eds), Essential oils and aromatic plants, ISBN-13: 978-94-009-5137-2

RESULTS AND DISCUSSION

Clary sage was used for the first experiments. Its oil and concrete are important in the flavour and fragrance industry. The concrete contains 38-45% of sclareol (3), a diterpenediol, that is used either as natural fixative or as raw material for the oxidative conversion to a mixture of ethers and unsaturated alcohols, possessing an ambergris-like aroma. In flavour industry, clary sage is used in alcoholic beverages, liqueurs and as a modifier in spices (4). Fresh plants were collected in Southern France and the essential oil was isolated from the fresh flowering tops either by hydrodistillation in a modified 'Karlsruher apparatus' (5) or by extraction with pentane-diethyl ether (7:3).

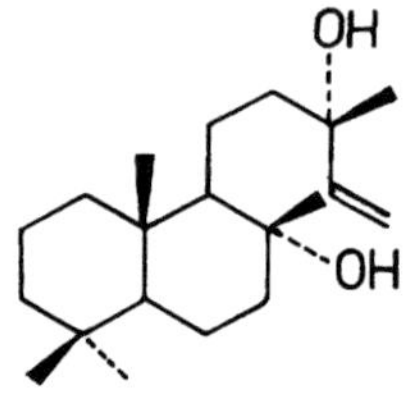

Significant differences between the essential oils obtained by hydrodistillation at pH 7.0 and by extraction of the plant material with the organic solvent mixture were shown by capillary GC. Only traces of sclareol were detected in the distilled oil, whereas it was the major constituent of the extract. Presumably, the reason for the relatively low volatility of this compound is its molecular weight (= 308.51). Linalyl acetate was identified as the second major constituent of the extract as well as of the hydrodistilled oil. However, the most striking difference between both samples was the occurrence of several additional main components in the distilled oil: myrcene, cis-β-ocimene, trans-β-ocimene, α-terpineol, nerol, geraniol, neryl acetate and geranyl acetate, while the amount of linalool was also drastically increased. We suspected linalyl acetate, respectively linalool formed by hydrolysis of the ester, to be the precursors of the artifacts, because of their lability (1, 2).

In order to correlate the artifacts mentioned to a certain precursor, and to get more detailed information about the rearrangement and decomposition processes, we investigated the influence of the acidity of the distillation medium on the stability of linalool and linalyl acetate in a model system by varying the pH over the range 2.2-8.0, using McIllvaine buffers (6) with pH values 2.2, 3.0, 4.0, 5.0, 6.0, 7.0 and 8.0 as described by Koedam (7). For the experiments carried out with linalyl acetate we used the apparatus shown in Fig. 1. A defined volume of the pure compound was refluxed with a buffer solution for 2 h. This should simulate the situation of the essential oil components during their iso-

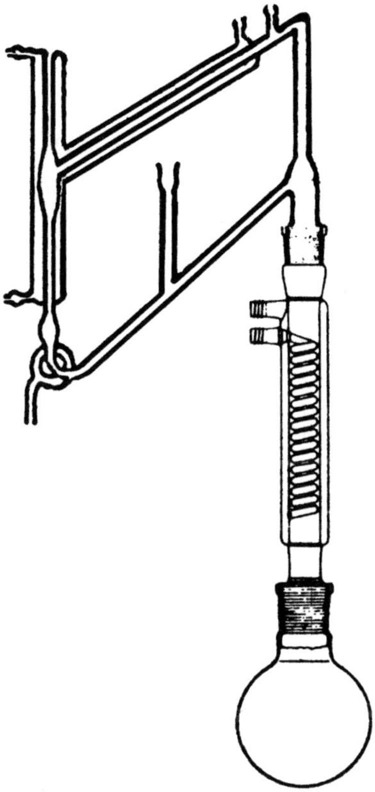

Fig. 1. Schematic drawing of the apparatus used to study the influence of the pH of the distillation medium on the stability of various compounds.

lation from the plant material. The gas chromatograms of the mixtures of components obtained from linalyl acetate at some selected pH values were compared with the chromatogram of the non-distilled reference. The most striking differences could be recognized at pH 2.2; they will be discussed later in connection with linalool. In the pH range 5.0-8.0 a similar pattern of artifacts was observed, and the components identified were in good accordance with the artifacts identified in the hydrodistilled oil of clary sage.

Subsequently we repeated the experiments with linalool instead of linalyl acetate. At pH 2.2 the pattern of the components was similar to that found for linalyl acetate, probably because hydrolysis of the ester is the first step in the reactions. However, geraniol refluxed under the same conditions yielded nearly the same mixture of artifacts. The reason might be a common intermediate (8). It is supposed (8) that geraniol forms a cation which is the resonance stabilized hybrid of the conformers geranyl, neryl and linalyl cation, and cyclization of the cations results in the formation of α-terpinyl cation (Fig. 2). This intermediate that probably arises from linalool, nerol and geraniol, or from their esters after hydrolysis, would explain the results obtained in our experiments.

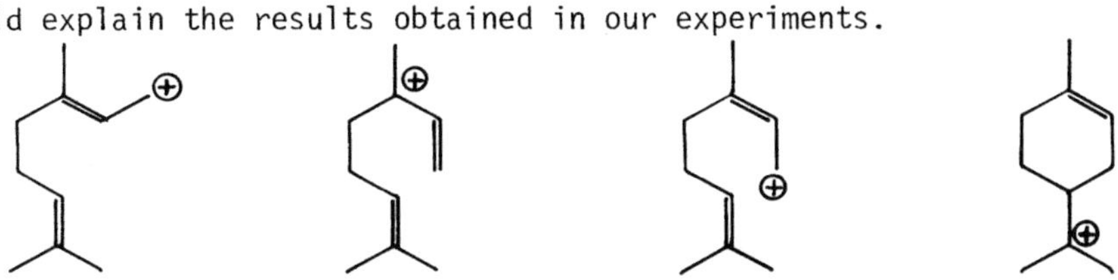

Fig. 2. Structural formulas of the intermediate cations (8). From the left to the right: geranyl, linalyl, neryl and α-terpinyl cation.

To get sufficient material for structure elucidation of the artifacts, a larger amount of linalool was refluxed at pH 2.2. Chromatographic and spectroscopic methods were used for identification of the GC peaks. In Table I the monoterpenes identified are listed. Linaloyl and ocimenoyl oxide were described for the first time as distillation artifacts in the essential oil of lime (9); they were probably formed as result of the low pH of the distillation water (organic acids in lime fruits).

Table I. Compounds identified after refluxing of linalool at pH 2.2 (? = tentatively identified).

Compound	Identification		
Myrcene	GC		
α-Terpinene	GC		
Limonene	GC		
cis-β-Ocimene	GC		
γ-Terpinene	GC		
trans-β-Ocimene	GC		
Terpinolene	GC		
Linaloyl oxide		MS	IR
1.4-Cineole		MS	IR
1.8-Cineole	GC	MS	IR
Ocimenoyl oxide		MS	IR
Linalool	GC		
p-Menth-3-en-1-ol		MS	
Terpinen-4-ol (?)	GC	MS	
β-Terpineol (cis or trans)		MS	
Myrcenol		MS	
cis-β-Ocimenol (?)		MS	
trans-β-Ocimenol (?)		MS	
α-Terpineol	GC	MS	
γ-Terpineol	GC	MS	

In our experiments we did not detect all the compounds mentioned, when clary sage and lavender oil were hydrodistilled from the plant material in non-buffered water, probably because the pH was not lower than 5.0. In contrast to linalyl acetate, linalool did not undergo rearrangements in the pH range 5.0-8.0. Therefore, it is obvious, that the artifacts found in the hydrodistilled oils from plant material are formed almost exclusively from linalyl acetate. Possibly the first reaction may be the formation of an ion pair with an organic acid molecule, which can undergo allylic rearrangement and degradation to yield the artifacts found (10).

Lavender, also containing linalool and linalyl acetate as major components of its essential oil, was collected in Southern France as well. Comparison of the essential oil isolated by hydrodistillation at pH 7.0 and the extract from the fresh plant material showed results similar to those described for clary sage. In the oil α-terpineol, neryl acetate, geranyl acetate, nerol and geraniol were identified as distillation artifacts. Differences in the group of monoterpene hydrocarbons were not so pronounced, since some of these compounds occur in the living plant, especially cis-β-ocimene and trans-β-ocimene.

It may be concluded that the analytical results obtained for an essential oil isolated by hydrodistillation should be interpreted carefully, and that solvent extraction is superior to distillation since it gives a more reliable picture of the genuine composition of an essential oil. However, since GC analysis of a plant extract leads to a faster deterioration of the GC column because of the non-volatiles also present in the extract, an effective sample preparation technique had to be found with the following criteria:

- the method should be simple, universally applicable, and able to yield sufficient amounts of volatiles for multiple GC, GC-MS or GC-FTIR;
- the method should prevent artifact formation, and yield the genuine (qualitatively and quantitatively) composition of the volatiles present in the living plant.

Finally a dynamic vapour phase stripping technique was chosen: the volatiles are isolated from the extract by a continuous stream of inert gas and subsequently trapped; a discrimination of essential oil constituents with a low vapour pressure should be avoided in that way. For an efficient trapping of the volatiles, the reflux trapping system with Freon 11, as described previously (11), seemed to have particular advantages. Fig. 3 shows a schematic drawing of the apparatus. A stream of purified Nitrogen (50 ml/min) passes through the concentrated extract for 40 h, and carries the volatiles to the extraction chamber. After concentration the sample is ready for GC analysis; the injection for GC is performed with a cooled micro-syringe.

Comparison of the gas chromatograms obtained for the lavender extract and for the sample prepared by gas phase stripping of the extract showed the efficiency of this technique. A good trapping efficiency of all volatiles, and only a slight discrimination of the constituents with a low vapour pressure, and no artifact formation were observed (Table II).

Table II. Percentage composition of lavender oil isolated by hydrodistillation at pH 7.0, by extraction, and by gas phase stripping of the extract (? = tentatively identified).

Compound	Hydrodistillation pH 7.0	Extraction	Extraction and gas phase stripping
Myrcene	1.90	0.48	0.59
Limonene	0.56	0.16	0.19
β-Phellandrene 1.8-Cineole	0.49	0.46	0.68
cis-β-Ocimene	4.53	4.45	5.99
trans-β-Ocimene	6.21	3.43	4.99
3-Octanone (?)	1.52	0.84	1.12
p-Cymene	0.72	0.34	0.47
Terpinolene	0.36	-	-
Oct-1-en-3-yl acetate (?)	1.85	1.72	2.09
cis-Linalool oxide	0.31	0.46	0.58
trans-Linalool oxide	0.18	0.30	0.38
Camphor	0.46	0.37	0.47
Linalool	40.91	24.95	25.78
Linalyl acetate	13.26	42.98	40.17
Terpinen-4-ol	2.36	2.81	2.45
β-Caryophyllene	3.96	5.48	4.72
α-Terpineol	5.43	-	-
Borneol α-Terpinyl acetate	2.15	1.66	1.48
Neryl acetate	1.22	-	-
Geranyl acetate	2.47	-	-
Nerol	1.04	-	-
Geraniol	3.24	-	-

The results obtained for clary sage* concerning the trapping efficiency etc were comparable with those for lavender. Even the relative amounts of sesquiterpenes like β-caryophyllene and germacrene-D were in good accordance with the extract, while sclareol had not been transported, due to its high molecular weight. According to a recent paper (12) only low and medium molecular weight compounds — up to eicosane in the alkane series — may be analyzed by gas phase stripping.

Summarizing it can be concluded that a combination of solvent extraction and gas phase stripping by means of a reflux trapping apparatus, yields an essential oil the composition of which is much more like that occurring in the living plant than that of the oil isolated by hydrodistillation.

*For illustration, see p. 134.

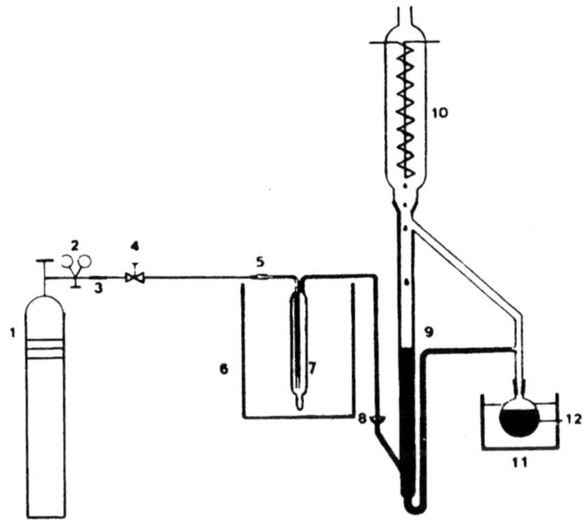

Fig. 3. Schematic drawing of the apparatus used to isolate the volatiles from plant extracts by gas phase stripping (10): 1. inert gas supply; 2. pressure reducer; 3. charcoal filter; 4. flow regulator; 5. connection with Teflon shrinking tube; 6. water bath (30°C); 7. sample container with plant extract; 8. connection by ground-ball-socket joint; 9. extraction chamber; 10. reflux condenser (-30°C); 11. water bath (30°C); 12. flask with Freon 11.

REFERENCES

(1) Crabalona LL. 1960, Soap Perfum. Cosmet. 33:495.
(2) Pickett JA, Coates J, Sharpe FR. 1975, in: Proceedings 15th Congress European Brewery Convention, Nice 1975.
(3) Arctander S. 1969, Perfume and Flavor Chemicals (Aroma Chemicals), Montclair (N.J., USA).
(4) Leffingwell JC, Stallings JW, Sellers FO, Lloyd RA, Kane FC. 1974, in: International Essential Oil Congress, San Francisco 1974.
(5) Sprecher E. 1963, Dtsch. Apoth.-Ztg. 103:213.
(6) Rauen HM. 1964, Biochemisches Taschenbuch, 2. Auflage, Berlin/Göttingen/Heidelberg, Springer-Verlag.
(7) Koedam A. 1980, Thesis, University of Leiden, The Netherlands.
(8) Stevens KL, Jurd L, Manners G. 1972, Tetrahedron 28:1939.
(9) Strickler H, Kováts E sz. 1966, Helv. Chim. Acta 49:2055.
(10) Kogami K, Kumanotani J. 1974, Bull. Chem. Soc. Japan 47:226.
(11) Rapp A, Knipser W. 1980, Chromatographia 13:698.
(12) Poole CF, Schuette SA. 1983, J. High Resol. Chromatogr. & Chromatogr. Commun. 6:526.

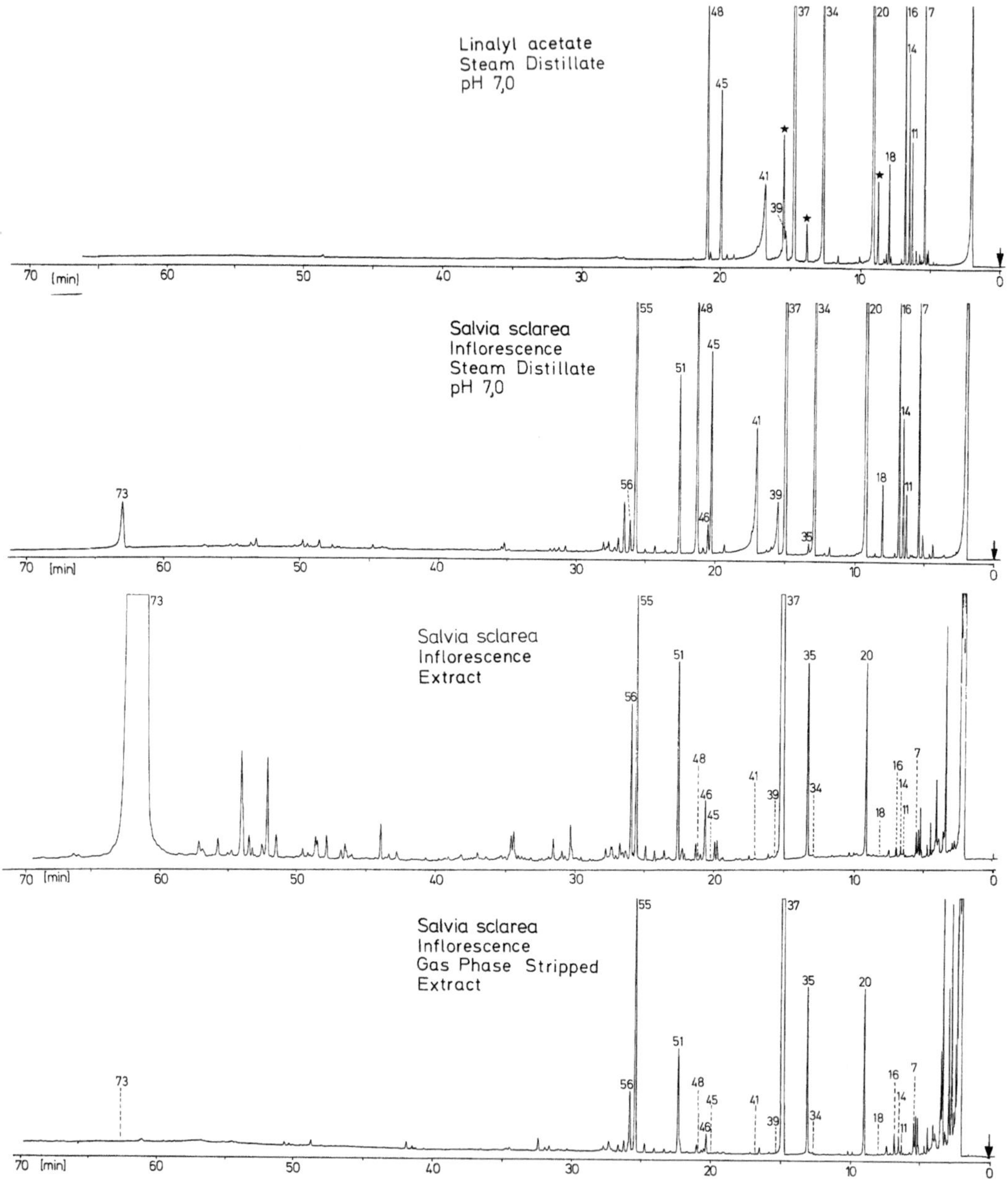

Fig. 4. Gas chromatograms of the hydrodistillate of linalyl acetate (pH 7.0) and of the essential oils of clary sage obtained by different methods. Column, capillary (30 m) coated with SE-30; oven temperature, 70-200°C (2.5°C/min); 7 = myrcene; 11 = limonene; 14 = cis-β-ocimene; 16 = trans-β-ocimene; 18 = terpinolene; 20 = linalool; 34 = α-terpineol; 37 = linalyl acetate; 39 = nerol; 41 = geraniol; 45 = neryl acetate; 48 = geranyl acetate; 73 = sclareol.

DETERMINATION OF THE ENANTIOMERIC COMPOSITION OF NATURAL FLAVOURING AGENTS BY ^{1}H-NMR SPECTROSCOPY

U. RAVID, E. PUTIEVSKY, V. WEINSTEIN, R. IKAN

ABSTRACT

The (+)- and (-)-enantiomers of linalool were isolated as the major monoterpene alcohols from the essential oils of coriander and sweet basil, respectively. The enantiomeric composition of natural (+)- and (-)-, and of synthetic (±)-linalool was examined by ^{1}H-NMR using a chiral lanthanide shift reagent.

INTRODUCTION

(+)-linalool is a major component in essential oils such as those of coriander, palmarosa, mace, petitgrain and sweet orange flowers. (-)-linalool is present in the oils of sweet basil, *Cinnamomum camphora* var. *orientalis*, *C. camphora* var. *occidentalis*, cajenne rosewood, neroli, linaloe, bergamot, lavandin and others. A mixture of (+)- and (-)-linalool has been reported in Brazil rosewood (1). "Odour" receptors in man and in insects can discriminate between enantiomers. In some cases one enantiomer can synergize the other (2).

Optical purity of enantiomers with known absolute configuration is of great importance in the chemistry of natural products, especially in the fields of flavours, fragrances and pheromones. The pure enantiomers of carvone show different odour properties (3, 4). Differences in the pheromonal activities on black-tailed deer and on khapra beetle were achieved by using highly pure enantiomers and mixtures of them (5, 6).

MATERIALS AND METHODS

Sweet basil (*Ocimum basilicum*, Mediterranean-type) and coriander (*Coriandrum sativum*) were grown in an experimental field at the Neve Ya'ar Agricultural Experiment Station. The plants were harvested in the summer of 1983. Fresh leaves of sweet basil were steam-distilled for 2 h in a 130 l direct steam pilot plant apparatus. Fresh fruits of coriander were hand separated and hydrodistilled in a modified Clevenger-type ap-

A. Baerheim Svendsen and J.J.C. Scheffer (eds), Essential oils and aromatic plants, ISBN-13: 978-94-009-5137-2

paratus for 2 h. The oils were dried over anhydrous sodium sulfate and stored at 2-6^{0}C. Linalool was isolated from both oils by fractional distillation on a Perkin-Elmer Model 131T Microstill at 0.1-0.2 mbar. Samples of 0.1 µl of the essential oils and of linalool were injected on a packed glass column (3 m x 4 mm i.d.) with 5% Carbowax 20M on Chromosorb W (80-100 mesh). Operating conditions: temperature program 80-200^{0}C (6^{0}C/min); carrier gas, Nitrogen, 30 ml/min. A Varian 3700 gas chromatograph equipped with FID, and a Hewlett-Packard 3390A integrator were used.

Optical rotations were measured on a Perkin-Elmer Model 531 polarimeter and the ^{1}H-NMR spectra recorded in deuterochloroform solutions on a Bruker WH 300 spectrometer.

RESULTS AND DISCUSSION

Linalool is the major component in the essential oils of sweet basil (63%) and coriander (77%). By fractional distillation of the oils in vacuo, linalool was isolated in high purity. The specific rotation of linalool isolated from sweet basil oil was -21.0^{0} (c = 1.00; ethanol), and from coriander oil +16.0^{0} (c = 1.54, ethanol). The absolute configuration of linalool was proved by pyrolytical formation from the diastereomeric pinan-2-ols (7) and by a seven-step synthesis from chiral 2,2,4-trimethyl-1,3-dioxolan-4-carbaldehyde (8).

Chiral lanthanide shift reagents were found to cause larger enantiomeric shift differences in the NMR spectrum of several chiral alcohols (2). $Eu(hfbc)_3$ reagent is paramagnetic and causes downfield changes in the chemical shift of the protons near the site of the β-diketone complex with the polar parts of the substrate without line broadening.

Addition of the natural (+)- and (-)-linalool and of synthetic (±)-linalool into a solution of $Eu(hfbc)_3$ in deuterochloroform in different concentrations cause downfield shifts of the protons of the complex. The absorption of the olefinic protons appearing about 7.5 ppm (by addition of 8.2 mg linalool to 40 mg $Eu(hfbc)_3$ in 0.4 ml deuterochloroform) is a doublet for (R)-(-)-linalool isolated from sweet basil oil. It shows that this linalool is a pure enantiomer. The absorption is split into two equal doublets by addition of synthetic (RS)-linalool to the chiral shift reagent. The enantiomeric ratio is 1:1. In the case of (S)-(+)-linalool isolated from coriander oil, each pair of the doublets is not equal. Spiking of the (S)-(+)-enantiomer with the pure (R)-(-)-enantiomer shows that the absorption at the lower field is that of the enantiomeric peak of (S)-(+)-linalool. The enantiomeric composition of linalool from the oil of coriander was found to be the same in both techniques of conventional polarimetry and of ^{1}H-NMR "polarimetry". The NMR spectra of synthetic and natural linalool in the presence of Eu(hfbc)$_3$ are shown in Fig. 1.

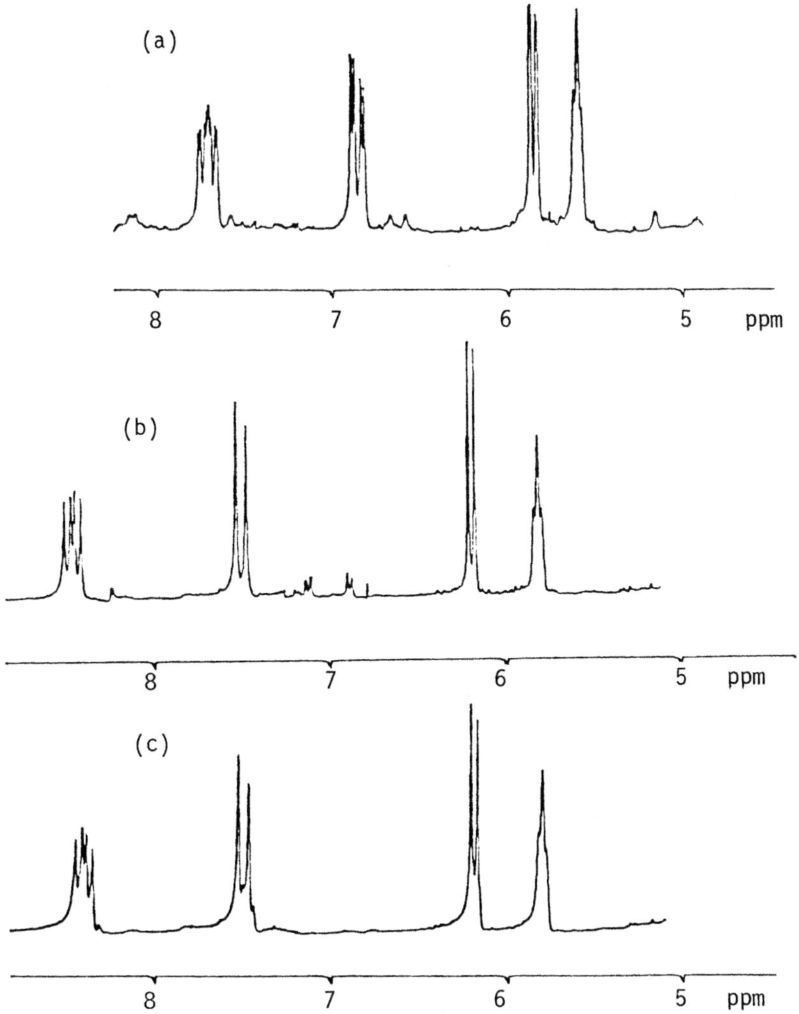

Fig. 1. NMR spectra of (a) synthetic racemic linalool, (b) natural linalool from sweet basil oil and (c) natural linalool from coriander oil, in presence of Eu(hfbc)$_3$.

REFERENCES

(1) Furia TE, Bellanca N, Eds. 1975, Fenaroli's Handbook of Flavour Ingredients, 2nd Ed., Vol.1, Cleveland, Ohio, CRC Press.
(2) Plummer EL, Stewart TE, Byrne K, Pearce GT, Silverstein RM. 1976, J. Chem. Ecol. 2:307.
(3) Russell GF, Hills JI. 1971, Science 172:1043.
(4) Friedman L, Miller JG. 1971, Science 172:1044.
(5) Müller-Schwarze D, Ravid U, Claesson A, Singer AG, Silverstein RM, Müller-Schwarze C, Volkman NJ, Zemanek KF, Butler RG. 1978, J. Chem. Ecol. 4:247.
(6) Greenblatt RE, Burkholder WE, Cross JH, Cassidy Jr. RF, Silverstein RM, Levinson AR, Levinson HZ, Ravid U. 1977, 174th Am. Chem. Soc. National Meeting, Chicago.
(7) Ohloff G, Klein E. 1962, Tetrahedron 18:37.
(8) Barner R von, Hübscher J. 1983, Helv. Chim. Acta 66:880.

DETERMINATION OF THE ENANTIOMERIC COMPOSITION OF SYNTHETIC FLAVOURING AGENTS BY ^{1}H-NMR SPECTROSCOPY

V. WEINSTEIN, R. IKAN, U. RAVID, E. PUTIEVSKY

ABSTRACT

A series of enantiomerically pure monocyclic monosubstituted γ-lactones was synthesized starting from optically active glutamic acid. The enantiomeric composition was examined by ^{1}H-NMR spectroscopy using the chiral solvation reagent [(R)-(-)-2,2,2-trifluoro-1-(9-anthryl)ethanol].

INTRODUCTION

The saturated five-membered lactone ring (butanolide) is the structural unit of numerous natural products possessing various biological activities. Many monocyclic substituted butanolides were isolated from natural sources. There are, however, only few synthetic methods for the preparation of various analogues of natural butanolides (1-5). Furthermore, the number of stereospecific routes leading to enantiomerically pure products is limited.

Optical activity of butanolides, as of many natural products, is essential for their biological activity (6-7). The purpose of the present study was to synthesize stereospecifically different series of optically active monosubstituted monocyclic butanolides, in order to evaluate their biological activity.

The optically active butanolides were prepared according to the following route (8):

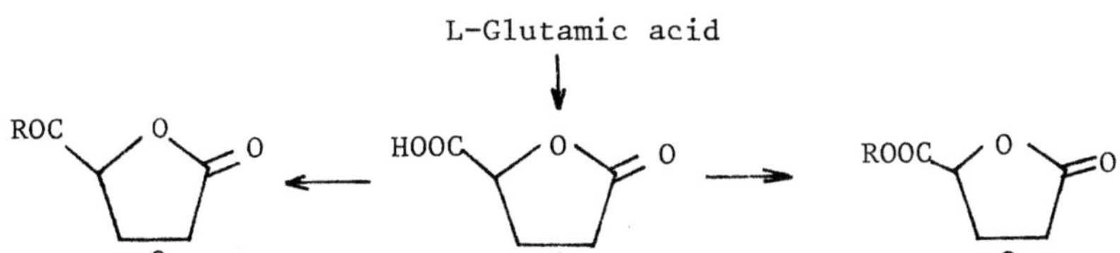

3 1 2

3a: R = alkyl

3b: R = R_1R_2N

A. Baerheim Svendsen and J.J.C. Scheffer (eds), Essential oils and aromatic plants, ISBN-13: 978-94-009-5137-2

MATERIALS AND METHODS

(S)-(+)-4-Carboxy-γ-butyrolactone 1 (see structural formula). Optically active L-glutamic acid (1 mole) was dissolved in water (400 ml); concentrated hydrochloric acid (205 ml) was added and the solution was cooled to 0-5°C. Sodium nitrite (1.5 mole) in water (220 ml) was added dropwise during a period of 5 h with magnetic stirring. The reaction mixture was left overnight at room temperature and then lyophilized. The yellow oily residue was crystallized from a mixture of ethyl acetate- petrol ether, yielding 52% of white crystals, m.p. 70-72°C, $[\alpha]_D^{20}$ = +10.6^{o}C (c = 5.0, ethanol).

4-Alkylaminocarbonyl-γ-butyrolactones 3b, general procedure. Compound 1 (1 mole) was gently refluxed with thionyl chloride (2 moles) for 2 h. Excess of thionyl chloride was distilled under reduced pressure yielding S-(+)-4-chlorocarbonyl-γ-butyrolactone (85%).

Alkylamine (0.2 mole) in dried diethyl ether (100 ml) was added dropwise with stirring to a cold solution of 4-chlorocarbonyl-γ-butyrolactone (0.1 mole) in dried diethyl ether (120 ml), and the mixture was stirred for 1.5 h. After filtration, the diethyl ether was distilled off and the residue crystallized from ethanol.

4-Alkylcarbonyl-γ-butyrolactones 3a were prepared by reaction of 4-chlorocarbonyl-γ-butyrolactone and alkyl bromides via alkyl cadmium compounds.

4-Alkoxycarbonyl-γ-butyrolactones 2 were prepared by reaction of 1 with various alcohols, using cation exchange resins as catalysts.

Optical rotations were measured on a Perkin-Elmer Model 531 polarimeter. ^{1}H-NMR spectra were recored in deuterochloroform solutions on a Bruker WH 300 spectrometer.

RESULTS AND DISCUSSION

The polarimetric method is widely used in determining the enantiomeric purity of many natural products. This technique could not be adopted in the present work due to two reasons:

- most of the products are new, and there is no external standard of enantiomeric purity;
- the contamination with minute quantities of optically active compounds may introduce a significant error in its value (9), since the optical rotation values of γ-butyrolactones are rather low.

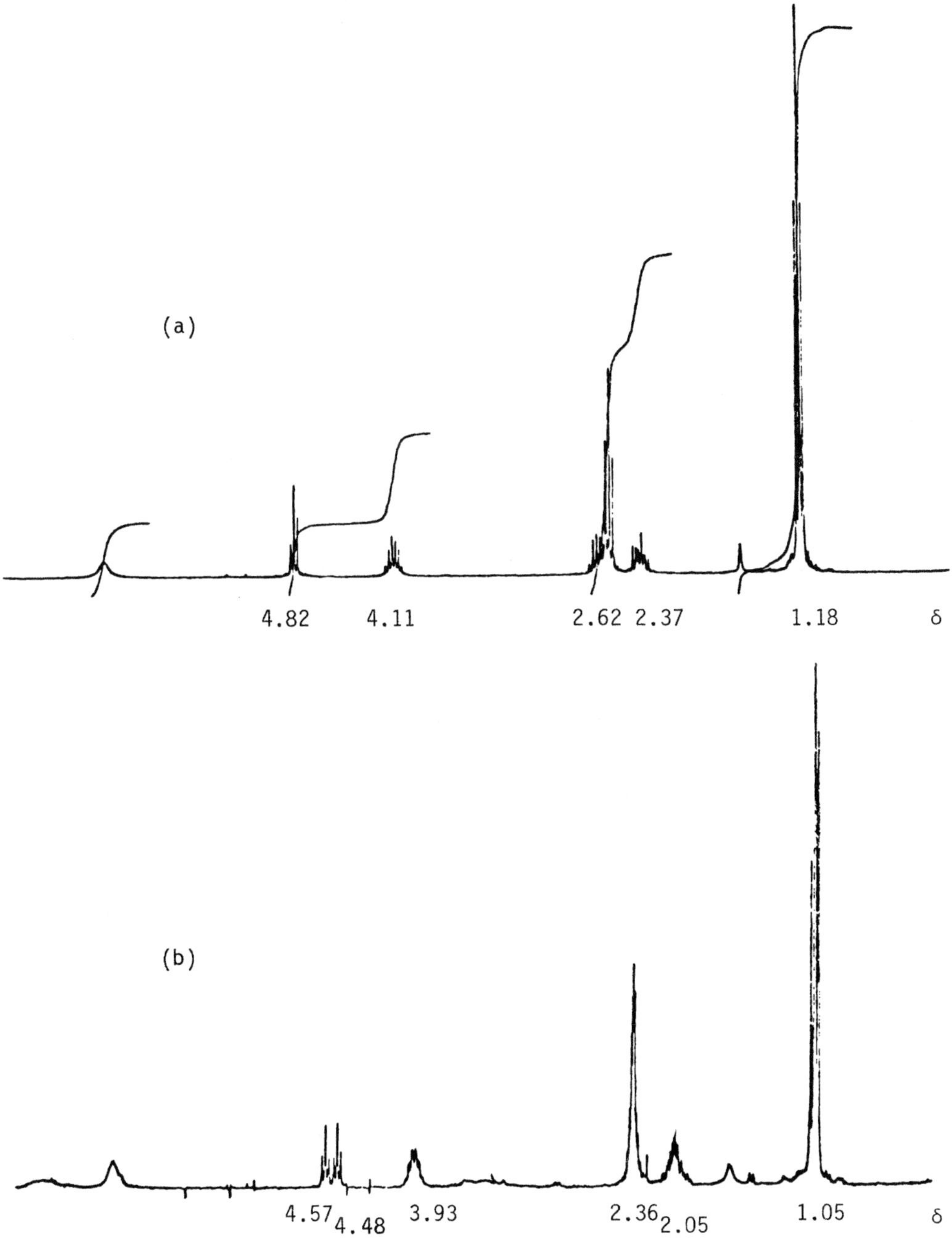

Fig. 1. NMR spectra of enantiomerically pure 4-isopropylaminocarbonyl-γ-butyrolactone. (a) neat; (b) in presence of TFAE

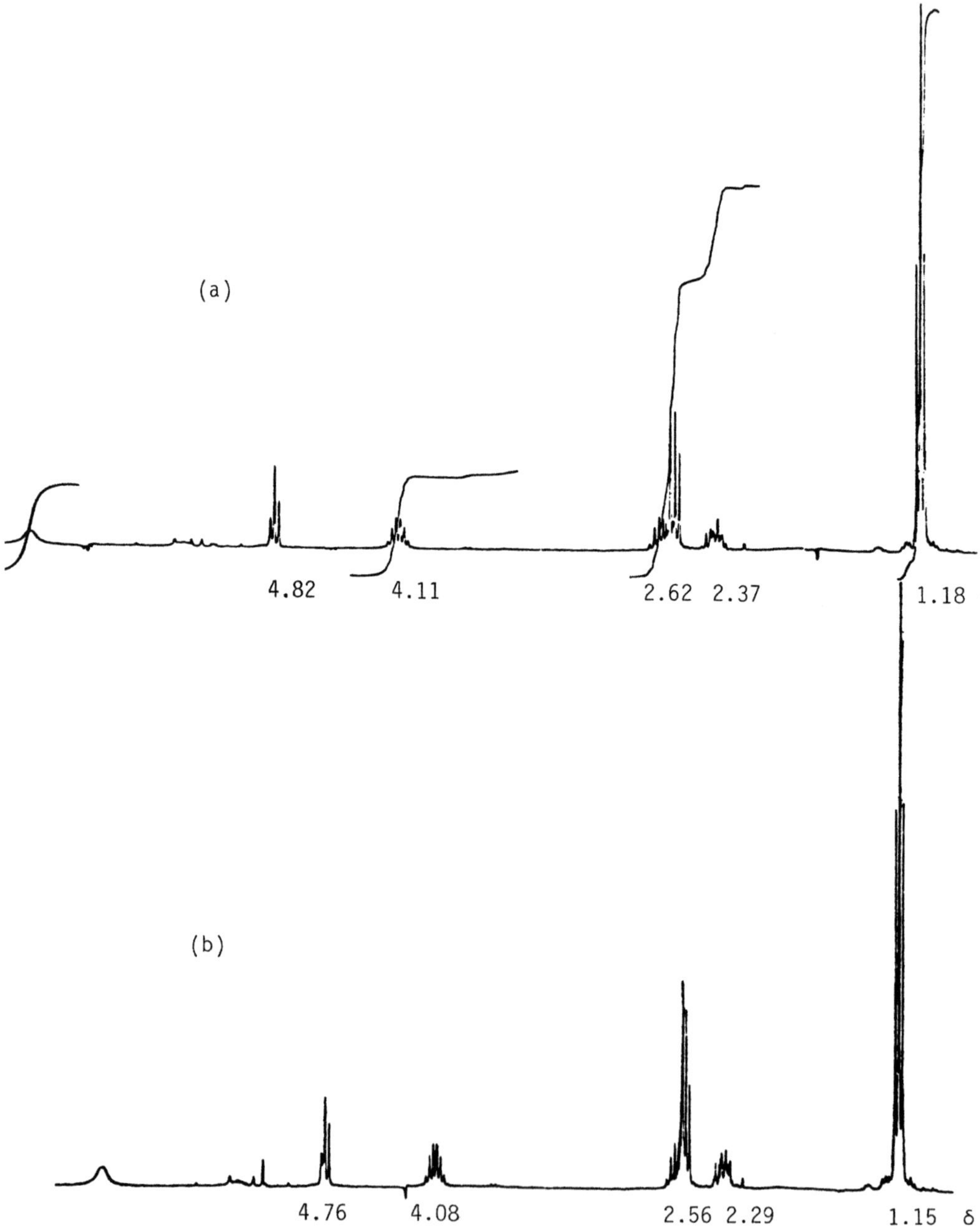

Fig. 2. NMR spectra of racemic 4-isopropylaminocarbonyl-γ-butyrolactone. (a) neat; (b) in presence of TFAE

It has been reported (10-12) that chiral (R)-(-)-2,2,2-trifluoro-1-(9-anthryl)ethanol (TFAE) induces spectral nonequivalence between enantiomeric lactones. Substituents on either face of the lactone respond differently to the shielding effect of the anthryl substituent of TFAE. The proton signals of the racemic mixture appear in the NMR spectrum as two splitted groups of peaks. NMR data of enantiomerically pure and racemic 4-isopropylaminocarbonyl-γ-butyrolactones (prepared by the same route from optically inactive glutamic acid) in the presence of TFAE may serve as an example for its useful application in the determination of enantiomeric purity of the substituted γ-lactones prepared by the routes outlined above (8) (see Figs. 1 and 2).

As expected, the NMR spectra of the racemic and optically active butanolides in absence of TFAE were absolutely identical. However, addition of the chiral solvation reagent caused splitting of peaks of the racemic γ-lactones, leaving the spectrum of the enantiomerically pure butanolide unchanged. The enantiomeric ratio in the racemic γ-lactone is 1:1. The nonequivalence of protons in different enantiomers is observed for the protons on the asymmetric carbon as well as for adjacent protons and the amide moiety of the 4-isopropylaminocarbonyl-γ-butyrolactone. For example, the proton on the asymmetric carbon of the enantiomerically pure γ-lactone is observed as one triplet at 4.76 ppm (J = 7.43 Hz), whereas the absorption of the analogous proton of the racemic mixture is split into two equal triplets: at 4.48 ppm (J = 7.43 Hz) and 4.57 ppm (J = 7.04, 7.43 Hz). The upfield shifts observed for all proton signals of the lactone depend on the molar ratio between the TFAE and the substrate (1).

REFERENCES

(1) Ravid U, Silverstein RM. 1977, Tetrahedron Lett. 5:423.
(2) Ravid U, Silverstein RM. Smith L. 1978, Tetrahedron 34:1449.
(3) Midland M, Tramontano A. 1980, Tetrahedron Lett. 21:3549.
(4) Jacovac I, Goodbrand H, Lok K. 1982, J. Am. Chem. Soc. 104:4659.
(5) Vigneron J, Blay V. 1980, Tetrahedron Lett. 21:1735.
(6) Boelens H. 1974, Proc. Internat. Symp. Food Sci. and Techn., Madrid, 1, p. 79.
(7) Riley R, Silverstein RM. 1974, Science 183:760.
(8) Weinstein V. 1983, Synthesis of γ-Lactones with Potential Biological Activity, Ph.D. Thesis, Hebrew University, Jerusalem.
(9) Hiyama T, Mishima T, Sawada H, Nozaki H. 1976, J. Am. Chem. Soc. 98: 641.
(10) Pirkle W, Hanske J. 1976, J. Org. Chem. 41:801.
(11) Pirkle W, Sikkenga D, Pavlin M. 1977, J. Org. Chem. 42:384.
(12) Pirkle W, Sikkenga D. 1977, J. Org. Chem. 42:1370.

CONSTITUENTS OF THE ESSENTIAL OIL OF *SALVIA STENOPHYLLA* — FIRST IDENTIFICATION OF (+)-EPI-α-BISABOLOL IN NATURE

E.-J. BRUNKE, F.-J. HAMMERSCHMIDT

ABSTRACT

By GC-MS 44 constituents of the essential oil of *Salvia stenophylla* were identified. Myrtenyl methyl ether could be detected for the second time and (+)-epi-α-bisabolol for the first time as naturally occurring substances. The sesquiterpene alcohol, α-bisabolol, can exist in four stereoisomeric forms, three of which have previously been discovered in nature. The missing stereoisomer, (+)-epi-α-bisabolol, was isolated from the essential oil of *S. stenophylla*.

INTRODUCTION

Around 1976 an essential oil called 'Essence Stenophylla' appeared on the market. It was not until 1980 that a brief description of its ingredients was published (1). Thanks to better analytical techniques, we succeeded in identifying some additional constituents and in revising some of the structures attributed by Garnero and co-workers (1).

Salvia stenophylla is well known in the folk medicine of Southern Africa. So, members of the more southerly Sotho-speaking cultures, now Lesotho and Botswana, burn the plant in their huts to disinfect them after illness and to drive away bugs. They also mix the plant with tobacco (2).

MATERIAL AND METHODS

The essential oil was obtained by steam distillation from dried *Salvia stenophylla* (Burch. ex) Benth. (Labiatae). *S. stenophylla* is a hardy, herbaceous, erect plant which grows wild in Southern Africa. Its quadrangular stem grows to 60 cm in height; it is covered with very short, stiff glandular hairs and a small number of spherical oil cells. Its leaves range from narrow and linear to elongated and lanceolate. They are about 5 x 1 cm, pinnatified to pinnate, with up to 10 pairs of finely serrated segments or lobes. They have a pubescence of very short glandular hairs, mostly on the veins of the underside of the leaf, and a large number of oil cells. The stem carries up to 10 whorls of inflorescence with 6-8 flowers each. These

A. Baerheim Svendsen and J.J.C. Scheffer (eds), Essential oils and aromatic plants, ISBN-13: 978-94-009-5137-2

whorls are standing 15-20 mm from each other low down on the stem, but they are closer near the top. The petals are pale-blue or pinkish. *S. stenophylla* flowers through the summer, i.e. from October to April. It grows on sandy or chalky soil, on riverbanks, and in damp habitats, as well as on open grass-land and savanna. The plant is found at an altitude of 400-2300 m (3-5).

The essential oil of *S. stenophylla* is a straw-coloured, mobile liquid. Its physical constants are:

d_4^{20} : 0.9075; n_D^{20} : 1.4860; $[\alpha]_D^{20}$: +3.8°.

The smell of the essential oil (on test strips) is fresh, herby, and camphoric. After a few minutes the odour changes to a fresh-spicy complex (pepper, clove, nutmeg), then after a couple of hours into a highly agreeable, herby-sweet and spicy fragrance (tea, lavender, cloves). After a day it leaves a slightly spicy, sweet, and somewhat minty afternote (hay, tea).

For gas chromatography-mass spectrometry a Hewlett-Packard 5992A was used (at 70 eV); column, glass capillary (25 m) WG-11; temperature program, 60-220°C (4°C/min); injector, 250°C; detector (FID), 250°C; carrier gas, Helium; injected amount, 0.1 µl (split 1:20).

RESULTS AND DISCUSSION

Fig. 1 shows a gas chromatogram of the essential oil of *S. stenophylla*, and Table I lists the constituents identified by GC-MS analysis of the whole, not fractionally distilled oil. Among a number of well known natural substances two are of special interest: myrtenyl methyl ether and α-bisabolol. The ether has only been found in hyssop oil (6).

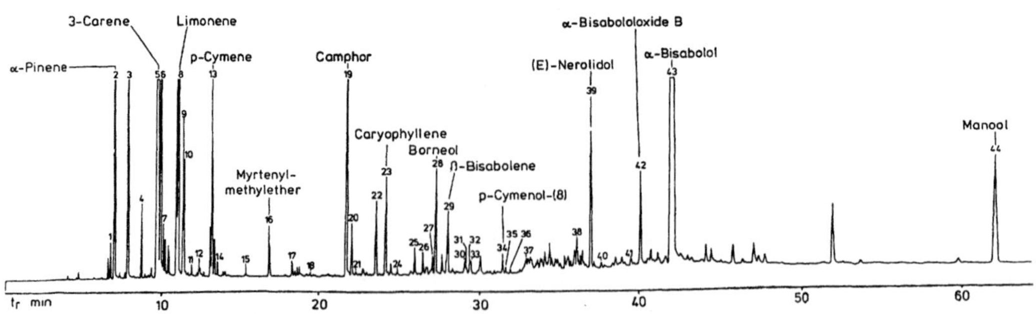

Fig. 1. Gas chromatogram of the essential oil of *Salvia stenophylla*.

Table I. Composition of the essential oil of *Salvia stenophylla*. The constituents were identified by GC-MS; for conditions, see Material and Methods; tr = trace (< 0.03%).

Peak No.	Compound	%
1	Tricyclene	0.2
2	α-Pinene	3.6
3	Camphene	3.4
4	β-Pinene	0.5
5	Δ3-Carene	20.0
6	Myrcene	6.2
7	α-Phellandrene	0.3
8	Limonene	3.6
9	β-Phellandrene	1.0
10	1.8-Cineole	0.7
11	cis-β-Ocimene	0.08
12	γ-Terpinene	0.1
13	p-Cymene	1.4
14	Terpinolene	0.1
15	6-Methylhept-5-en-2-one	0.1
16	Myrtenyl methyl ether	0.4
17	4-Isopropenyltoluene	0.1
18	trans-Sabinene hydrate	tr
19	Camphor	3.3
20	Linalool	0.4
21	cis-Sabinene hydrate	0.03
22	Bergamotene	0.7
23	β-Caryophyllene	1.0
24	Myrtenal	0.06
25	E-β-Farnesene	0.2
26	α-Humulene	0.2
27	α-Terpineol	0.2
28	Borneol	1.0
29	β-Bisabolene	0.6
30	δ-Cadinene	0.1
31	γ-Cadinene	0.09
32	ar-Curcumene	0.1
33	Myrtenol	0.1
34	p-Cymen-8-ol	0.2
35	Calamenene	tr
36	Geranyl acetone	tr
37	α-Calacorene	0.1
38	Caryophyllene epoxide	0.3
39	E-Nerolidol	1.5
40	Humulene epoxide II	0.06
41	Spathulenol	0.1
42	α-Bisabolol oxide B	1.1
43	α-Bisabolol	29.8
44	Manool	2.9

Determination of the structure of (+)-epi-α-bisabolol, 4

Having two chiral centers, α-bisabolol has four possible stereoisomers, three of which — compounds 1, 2 and 3 — have previously been discovered in nature (7). The structure of (-)-α-bisabolol, 1, the principal active substance of chamomile oil, was found by X-ray structural analysis of a crystalline derivative and by partial synthesis, using (-)-(S)-limonene as starting material (8-11).

Babin et al. (9) synthesized the epimers 1 and 2 by prenylation of (-)-limonene, and the epimers 3 and 4 from (+)-limonene, both with retention of configuration on the ring (Fig. 2). The epimers were separated by chromatography of their p-nitrobenzoates. These authors (9) adopted a nomenclature based on 'sesquimenthane' numbering and the R,S-nomenclature for the characterization of the stereochemistry. We are using the same nomenclatures in this paper. The stereoisomeric bisabolols 1-4 cannot be distinguished by their GC retention behaviour on the achiral columns in general use today, or by their mass spectra. Thus for determination of their relative or absolute configurations, the α-bisabolol stereoisomers have to be isolated from essential oils or similar mixtures.

The relative configuration of the isomers on C_4 and C_8 can be determined by ^{1}H-NMR spectroscopy (9, 10). For (-)-α- and (+)-α-bisabolol, 1 and 3, the tertiary methyl group on C_8 gives a singlet at δ = 1.10 ppm, while for (-)-epi-α- and (+)-epi-α-bisabolol, 2 and 4, this signal appears at δ = 1.13 ppm. The relative configuration of the bisabolols can also be demonstrated by ^{13}C-NMR spectroscopy (Table II).

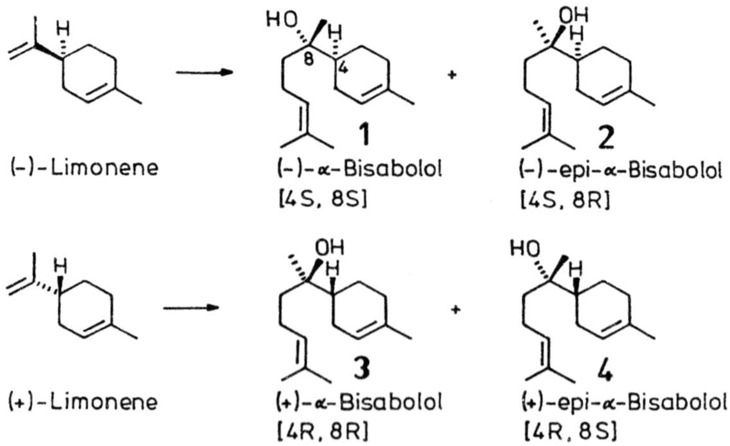

Fig. 2. Synthesis of bisabolol epimers (9).

Table II. ^{13}C-NMR data of the bisabolols. A = data obtained in the present study; B = data from the literature (7).

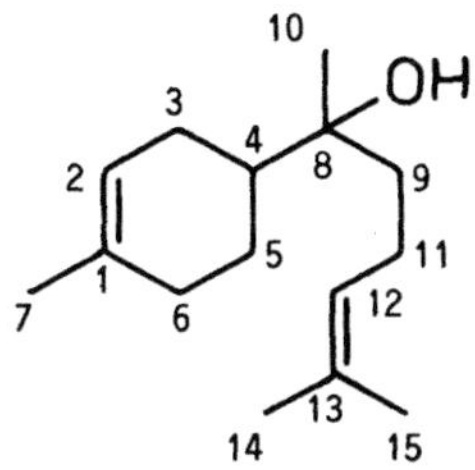

α-Bisabolol A	α-Bisabolol B	Multiplicity	C-atom	epi-α-Bisabolol A	epi-α-Bisabolol B
17.66	17.65	q	C-15	17.63	17.64
22.20	22.23	t	C-11	22.42	22.39
23.20	23.33	q	C-7 ?	23.32	22.82
23.35		q	C-10	23.90	23.33
23.35	23.45	t	C-5	24.18	24.12
25.69	25.68	q	C-14	25.69	25.69
27.05	27.04	t	C-3	26.23	26.21
31.17	31.19	t	C-6	31.23	31.19
40.32	40.32	t	C-9	39.62	39.55
43.16	43.21	d	C-4	43.53	43.50
74.12	74.15	s	C-8	74.09	74.22
120.80	120.70	d	C-2	121.10	121.00
124.92	124.91	d	C-12	124.98	124.85
131.22	131.32	s	C-13	131.14	131.44
133.86	133.91	s	C-1	133.46	133.64

As shown in Table III, the epimer pairs 1-2 on one hand and 3-4 on the other, each give very similar specific rotations. Thus it is better to arrive at the absolute configurations by a combination of NMR data and specific rotations.

The sesquiterpene alcohol which goes to make up some 30% of the essential oil of *S. stenophylla*, gives a mass spectrum almost identical to that of (-)-α-bisabolol. For determination of its relative and absolute configuration the compound was isolated by fractional distillation of 1 kg of the essential oil — supplied by Cavallier Frêres, Grasse, France — on a 1 m glass-packed column. The ^{1}H-NMR spectrum of the α-bisabolol isomer isolated showed the 8-epi form to be present. Since the compound, indicated by GC as 99.9% pure, was dextrorotatory, $[\alpha]_D^{20} = +67.4^o$, it had to be (+)-epi-α-bisabolol, 4. Thus, all four stereoisomers of α-bisabolol have now been shown to occur in nature.

Table III. Configurations, specific rotations and natural occurrences of the α-bisabolols. (*) = present study.

Stereoisomer	Configuration	$[\alpha]_D^{20}$	Natural occurrence
(-)-α-bisabolol, 1	4S,8S	-68.4° (9) -57.7° (7)	*Matricaria chamomilla* *Vanillosmopsis erythropappa*
(-)-epi-α-bisabolol, 2	4S,8R	-68.9° (7)	*Myoporum crassifolium* (12, 13)
(+)-α-bisabolol, 3	4R,8R	+54.9° (7) +51.7° (14) +53.8° (8)	*Populus balsamifera* (buds) *Atalantia monophylla correa*
(+)-epi-α-bisabolol, 4	4R,8S	+67.4° (*)	*Salvia stenophylla* (*)

ACKNOWLEDGEMENTS

We like to express our thanks to Prof.Dr. A. Nahrstedt, Technical University of Braunschweig, FRG for his kind assistance on botanical aspects, to Mr. W. Rojahn for the distillation of the oil, to Prof.Dr. H.M.R. Hoffmann, University of Hannover, FRG for the NMR analyses, and to Mrs. B. Kohlenberg and Mrs. A. Lückhoff for recording the mass spectra.

REFERENCES

(1) Jequier D, Nicollier G, Tabacchi R, Garnero J. 1980, Phytochemistry 19:461.
(2) Watt JM, Breyer-Brandwijk MG. 1962, The Medicinal and Poisonous Plants of Southern and Eastern Africa, 2nd edition, Edinburgh/London, E. & S. Livingstone Ltd., p. 527.
(3) Bentham G. 1832-1836, Labiatarum Genera et Species, London, James Ridgway & Sons, p. 306.
(4) De Candolle A. 1848, Prodomus Systematis Naturalis Regni Vegetabilis, Part 12, Paris, Victor Masson, p. 353.
(5) Hedge IC. 1974, Notes R. Bot. Gard. Edinburgh 33,No.1:77.
(6) Joulain D. 1976, Rivista Italiana E.P.P.O.S. 58:479.
(7) Flaskamp E, Nonnenmacher G, Isaac O. 1981, Z. Naturforsch. 36b:114; correction, Ibid. 36b:526.
(8) Prangée Th, Babin D, Fourneron J-D, Julia M. 1979, C. R. Acad. Sci. Paris Ser. C 383.
(9) Babin D, Fourneron J-D, Julia M. 1981, Tetrahedron 37:1.
(10) Schwartz MA, Swanson GC. 1979, J. Org. Chem. 44:953.
(11) Iwashita T, Kusumi T, Kakisawa H. 1979, Chem. Lett. 947.
(12) O'Brien KG, Penfold AR, Werner RL. 1953, Austr. J. Chem. 6:166.
(13) O'Brien KG, Penfold AR, Sutherland MD, Werner RL. 1954, Austr. J. Chem. 7:298.
(14) Sorm F, Vrany M, Herout V. 1952, Chem. Listy 46:364.

GC-MS ANALYSIS OF THE ESSENTIAL OIL OF *MERIANDRA BENGHALENSIS*

M. BRUNO, G. MELLERIO, F. PIOZZI, P. VITA-FINZI

ABSTRACT

Meriandra benghalensis (Roxb.) Benth. (Labiatae) is a leafy shrub known since a long time to give an essential oil rich in (+)-camphor. By GC-MS we could identify seven monoterpene hydrocarbons and seventeen oxygenated monoterpenes; the sesquiterpene content was very low: traces of three hydrocarbons, two alcohols (MW 222) and some unknown compounds (MW 220).

INTRODUCTION

Meriandra benghalensis (Roxb.) Benth. (Labiatae) is a leafy shrub (1-1.5 m height) growing on the highlands of the Horn of Africa and in India. The species was sometimes named *Salvia benghalensis* Roxb., *S. abyssinica* R. Brown and *S. schimperiana* Hoechst ex Benth.; it is acclimatized since 60 years in the Botanic Garden of the University of Palermo (Italy).

The plant was known since a long time to give an essential oil rich in (+)-camphor (1-3); experiments have been performed for an industrial isolation of this compound, but they never attained to production. The leaves are widely used as food aromatizer and in popular medicine, in the same manner as sage leaves. Besides the essential oil, the aerial parts contain triterpenes, flavones and several new sesquiterpenes (4). Older studies (5) on the essential oil suggested the occurrence of small amounts of 1.8-cineole and camphene, in addition to camhor, the main component.

We report here on the composition of the essential oil of *M. benghalensis*, obtained from fresh leaves harvested in Palermo.

MATERIAL AND METHODS

Fresh leaves of *M. benghalensis* (1200 g), cultivated in the Botanic Garden of Palermo and harvested in November 1983, were minced and subjected to steam distillation. The yellowish oil containing white crystals of (+)-camphor, was separated by means of a pipette, dried over anhydrous sodium sulfate and kept in a sealed vial under Nitrogen; yield 4.5 g

A. Baerheim Svendsen and J.J.C. Scheffer (eds), Essential oils and aromatic plants, ISBN-13: 978-94-009-5137-2

(0.375%).

Fresh leaves collected in June 1980 gave a somewhat higher percentage of essential oil and of crystals of camphor (0.45%); part of the crystals was filtered from the aqueous layer. The content of camphor was estimated as 82% of the oil.

(+)-Camphor was identified by m.p., IR, $[\alpha]_D$, and preparation of derivatives (semicarbazone, 2,4-dinitrophenylhydrazone).

The oil (4.5 g) was brought onto a column of silica gel (50 g) mixed with cyclohexane; three fractions were eluted to separate hydrocarbons from oxygenated compounds: A) 200 ml cyclohexane, 0.7 g; B) 200 ml cyclohexane-ethyl acetate 85:15, 3.5 g; C) 200 ml ethyl acetate, 0.2 g.

The essential oil and the fractions were subjected to GC-MS using a gas chromatograph Varian 2700 connected (glass jet) with a mass spectrometer DuPont 21492B equipped with a data system Hewlett-Packard 2109A. Packed columns:

- silanized glass (3 m x 2 mm i.d.), packed with 2% OV-225 on Gas Chrom Q (60-80 mesh); column temperature, 50-150°C (4°C/min) and then 150-250°C (8°C/min); injector, 250°C; interface (line and jet), 250°C; MS source, 210°C; carrier gas, Helium, 15 ml/min;
- silanized glass (1.8 m x 2 mm i.d.), packed with 5% SP-2100 + 0.1% SP-401 on Supelcoport (100-120 mesh); column temperature, 60-140°C (4°C/min) and then 140-210°C (12°C/min); injector, 220°C; interface, 240°C; MS source, 250°C; carrier gas, Helium, 30 ml/min.

Capillary GC analyses were run on a GC-MS system Finnigan-MAT 1020B. Column: fused silica (25 m x 0.30 mm i.d.) coated with OV-1 (film thickness 1-1.5 μm); column temperature, 40-250°C (6°C/min); injector, 250°C; interface, 250°C; MS source, 70°C; carrier gas, Helium, 1 ml/min.

RESULTS AND DISCUSSION

The terpenes identified are listed in Table I according to increasing retention times on the SP-2100 + SP-401 GC column, and their occurrence in the different chromatographic fractions. As above, (+)-camphor is largely the main constituent of the essential oil of *M. benghalensis*. Therefore we considered it useless to give the relative percentages of the other, minor, constituents; their relative amounts are indicated by crosses.

The mass spectra recorded were compared by computerized methods with

Table I. Components detected in the essential oil of *Meriandra benghalensis* using an SP-2100 + SP-401 GC column. a = identified on an OV-225 GC column; A, B and C = fractions collected by column chromatography; for details, see Material and Methods; tr = trace.

			GC columns				
	Molecular		Packed				Capillary
Compound	weight	Amount	Oil	A	B	C	Oil
Tricyclene	136	+	x	x			x
α-Pinene	136	++	x	x			x
Camphene	136	+++	x	x			x
β-Pinene	136	++	x	x			x
Myrcene (a)	136	tr	x	x			
Limonene (a)	136	tr					x
p-Cymene	134	tr	x				x
1.8-Cineole	154	++	x	x	x	x	x
No terpene (lactone?)		tr	x		x		
Camphenilone	138	tr	x				x
α-Campholene aldehyde (a)	152	tr					x
Camphor	152		x	x	x	x	x
p-Cymen-8-ol (a)	150	tr					x
Carveol	152	tr				x	
Borneol	154	++	x		x	x	
Terpinen-4-ol	154	+	x		x	x	x
Myrtenal	150	+	x	x			
α-Terpineol	154	+	x		x	x	
Myrtenol (a)	152	tr			x	x	
Verbenone (a)	150	+	x				
Unknown	152	tr			x		x
Cumin aldehyde	148	tr		x			
Bornyl acetate	196	+	x	x			x
Carvone	150	+	x		x		
Piperitone	152	+	x		x		
Isobornyl acetate	196	tr	x				x
α-Gurjunene	204	tr		x			
β-Caryophyllene	204	tr		x			
Humulene	204	tr		x			
Unknown	220	+	x	x	x		x
Unknown	220	+	x	x	x		x
Unknown	220	+	x				x
Unknown	220	+	x	x			x
Unknown	220	++	x		x		
Unknown	220				x		
Caryophyllene oxide	220	tr			x		x
Palustrol	222	++	x	x	x		x
β-Eudesmol	222	+++	x		x		x
Unknown	218	tr		x			
Unknown	234				x		x
Unknown	220			x			
Hydroxycalamenene (?)	218			x			x

those present in our own files and with those in commercial libraries (6, 7). The spectra were also compared with those in the literature. In some cases we could compare the GC retention times with those of authentic samples.

It is noteworthy that the main monoterpenes are strictly related to camphor (camphene, borneol and its acetate, camphenilone and campholene aldehyde) or are bicyclic monoterpenes of the pinane group (α- and β-pinene, myrtenol, myrtenal, verbenol and verbenone). However, many other monocyclic monoterpenes were identified.

The oxygenated derivatives (alcohols and carbonyl compounds) seem to be characteristic for the allylic oxydation.

Sesquiterpene hydrocarbons are almost absent. Only β-caryophyllene, humulene and α-gurjunene were identified. Among the alcohols only β-eudesmol and palustrol (or a strictly related alcohol) could be identified so far.

REFERENCES

(1) Zucco G. 1923, Boll. R. Soc. Geogr. Ital. Ser. V 12,No.5-6:268.
(2) Cavara F. 1924, Riv. Ital. Essenze Profumi (Milano) 6:13.
(3) Bruno F. 1926, Boll. Studi Inform. Giardino Coloniale Palermo 9:3.
(4) Perales A, Martinez-Ripoll M, Fayos J, Savona G, Bruno M, Rodriguez B. 1983, J. Org. Chem. 48:5318.
(5) Palazzo FC, Alinari E. 1924, Bolletino Informazioni Economiche Ministero Colonie (Roma) 12:388.
(6) Stenhagen E, Abrahamsson S, McLafferty FW. 1969, Atlas of Mass Spectral Data, New York, Interscience.
(7) Heller SR, Milne GWA. 1980, EPA/NIH Mass Spectral Data Base, Washington, NBS.

ESSENTIAL OILS OF ISRAELI WILD SPECIES OF LABIATAE*

U. RAVID, E. PUTIEVSKY

ABSTRACT

The essential oils of six wild Israeli species of the Labiatae family were studied. The main components in the oils of *Majorana syriaca*, *Coridothymus capitatus*, *Satureja thymbra*, *Salvia fruticosa*, *Salvia dominica* and *Micromeria fruticosa* were identified. Two phenolic chemotypes, one containing thymol, the other one carvacrol, were found in *Majorana syriaca* and *Coridothymus capitatus*. The hydrodistilled oils were analyzed by means of GLC and GC-MS. Identification of individual compounds isolated from the oils by fractional distillation and preparative GC was carried out by IR spectroscopy.

INTRODUCTION

Many species of the Labiatae family are aromatic and often used as herbs and spices in folk medicine, and as a source of fragrant chemicals (1, 2). There are 33 genera of the Labiatae, including 180 species, growing wild in Israel (3). Some of them are collected by local farmers. The phenol-containing species, *Majorana syriaca*, *Coridothymus capitatus* and *Satureja thymbra*, are popular seasonings in the Middle East (4). Leaves of *Salvia fruticosa* and *Micromeria fruticosa* are used in herbal and medicinal tea preparations (3, 5, 6). On the world market *Salvia fruticosa* is used as a substitute or additive to *Salvia officinalis*. The essential oil distilled from the inflorescence of *Salvia dominica* has a pleasant aroma and is rich in monoterpene alcohols.

MATERIALS AND METHODS

Plant material: During the years 1979-1983 a few hundred plants of *Majorana syriaca*, *Coridothymus capitatus*, *Satureja thymbra*, *Salvia fruticosa*, *Salvia dominica* and *Micromeria fruticosa* were transferred from several wild populations in Central and Northern Israel to an experimental

* Contribution from the Agricultural Research Organization, The Volcani Center, Bet Dagan, Israel. No. 1161-E, 1984 Series.

A. Baerheim Svendsen and J.J.C. Scheffer (eds), Essential oils and aromatic plants, ISBN-13: 978-94-009-5137-2

field at the Newe Ya'ar Agricultural Experiment Station. Seeds of a commercial variety of *Salvia sclarea* received from England were sown in rows in the autumn of 1982 and the mature inflorescences were distilled in the summer of 1983.

Isolation of the essential oil: the fresh plant material was steam-distilled for 2.5 h in a 130 l direct steam pilot plant apparatus. The isolated oil was dried over anhydrous sodium sulfate and stored at 4-6°C.

Gas liquid chromatography: the analyses were carried out using a Varian 3700 gas chromatograph equipped with FID, and a Hewlet-Packard 3390A integrator. Fractions were collected in a modified Brownlee-Silverstein thermal gradient collector. Two packed columns (3 m x 4 mm i.d.) were used, one packed with 5 % Carbowax 20M on acid washed, silanized Chromosorb W (80-100 mesh), the other with 6 % OV-17 on acid washed, silanized Chromosorb W (60-80 mesh). Operating conditions: isothermal 70°C, 2 min, then 70-210°C (6°C/min) and isothermal 3 min; carrier gas, Nitrogen, 30 ml/min; injection volume, 0.1 µl.

Identification of components: retention times of individual components were compared with those of authentic samples on both columns. GC-MS was performed using a computerized Finnigan 4000 system and fused silica capillary columns (25 m) coated with OV-1 and C-Wax 20M; carrier gas, Helium, 5 ml/min; IR spectra were obtained from samples dissolved in carbontetrachloride, using a Perkin-Elmer 298 apparatus.

RESULTS AND DISCUSSION

The six potentially valuable species are native perennial plants in the hilly regions of Northern Israel (7). Two species, *Majorana syriaca* and *Salvia fruticosa*, collected from wild populations or cultivated, are available commercially. The six species can be divided into three groups: the phenol-containing species, the *Salvia* species, and *Micromeria fruticosa*.

About 20 components, representing 94.3-98.8 % of the oils, were identified.

The phenol-containing species (*Majorana syriaca*, *Coridothymus capitatus* and *Satureja thymbra*) are collected in the wild by local farmers when they start to flower. The dried leaves, called "zaatar" in Arabic, together with salt, sesame seeds and fruits of the tree *Rhus coriaria*, constitute a popular seasoning in the Middle East (4).

Two chemotypes, one containing thymol and the other carvacrol, were found in *M. syriaca* and in *C. capitatus*; in *S. thymbra* only the carvacrol-type was found. (Table I). The two biosynthetic precursors of the phenols, γ-terpinene and p-cymene (8), were also found in high concentration in the three species. In the late growth stage, the concentration of carvacrol (59.4%) in *S. thymbra* was higher than in the early growth stage (11.8%). Young leaves of *M. syriaca* are also known to contain smaller amounts of thymol and carvacrol than older leaves (2). In the thymol-type of *C. capitatus* the concentration of thymol during the late growth stage (52.6%) was higher than in the early stage (39.3%). The highest phenol content was found in *M. syriaca*: 56.7% in the thymol-type and 52.6% in the carvacrol-type. *S. thymbra* contained the highest amount of the major monoterpene hydrocarbons, γ-terpinene and p-cymene (58.7%).

Table I. Percentage of the components of the essential oil from three phenol-containing species in an early stage of growth. tr = trace (< 0.1%).

Compound	*Majorana syriaca*		*Coridothymus capitatus*		*Satureja thymbra*
	thymol -type	carvacrol -type	thymol -type	carvacrol -type	carvacrol -type
α-Pinene	2.3	3.4	2.9	2.6	4.2
Camphene	0.1	0.2	0.3	1.2	0.3
β-Pinene	0.2	0.2	0.3	0.2	0.5
Sabinene	tr	tr	0.2	tr	0.3
Δ3-Carene	0.1	0.2	0.1	tr	0.1
Myrcene	2.9	2.7	3.0	2.2	3.3
α-Terpinene	2.8	2.5	3.2	1.9	3.7
Limonene	0.4	0.2	0.7	0.2	0.7
1.8-Cineole	0.3	0.2	0.6	0.6	0.3
γ-Terpinene	14.5	13.0	19.4	14.7	45.8
p-Cymene	13.1	19.7	6.0	13.6	12.9
Camphor	tr	0.5	tr	0.5	0.1
Linalool	0.2	tr	1.0	2.5	0.6
Terpinen-4-ol	0.6	1.0	0.8	3.0	0.4
β-Caryophyllene	2.2	1.6	5.0	5.1	5.3
Borneol	0.7	0.2	1.2	2.0	0.5
Thymol	41.1	12.4	39.3	8.1	6.7
Carvacrol	15.6	40.2	12.7	34.8	11.8
	97.1	98.2	96.7	93.2	97.5

Over 70% of the species of *Origanum* are found in the Mediterranean basin (9). Brieskorn and Brunner (10) reported that the main components in *O. maru* (= *Majorana syriaca*) were carvacrol (43.7%), thymol (30.9%) and p-cymene (11.3%). Zaitschik and Levontin (11) found that thymol was the main component in the oil, and Fleischer et al. (12) reported the existence of two phenolic chemotypes.

Origanum majorana from Turkey contains 65.1% carvacrol and 3.9% thymol (13), while *O. smyrnaeum* of Turkish origin contains carvacrol (72% and 64%) and thymol (0.4% and 0.6%) (14). Calzolari et al. (15) reported that Turkish origanum oil contained 83.1% carvacrol and 0.9% thymol. 73.6% carvacrol and 0.4% thymol were found in *Coridothymus capitatus* of Greek origin (16).

By choosing the right chemotype or cutting leaves of different ages, it is possible to control the organoleptic characters and the content of phenols of commercial origanum, thyme and savory essential oils.

Salvia is a large genus containing about 900 species. Among the *Salvia* species growing wild in Israel, *S. fruticosa* (= *S. triloba*) and *S. dominica* are of great importance as potential sources for commercial herbs and essential oils. *S. fruticosa* is a native shrub in the Mediterranean basin, 1-1.5 m high, which flowers from March till June. Tea prepared from fresh or dried leaves is commonly used as a remedy for stomach pains, coughs and colds (3, 5, 6). The main components in the oil distilled from leaves are 1.8-cineole, camphor, α-pinene, β-pinene, borneol and β-caryophyllene. The amounts of α- and β-thujone, the major components in *S. officinalis*, were below 0.1%. The proportion of the monoterpene hydrocarbons in the oil was 28.4% (Table II).

Table II. Composition of the essential oil from *Salvia fruticosa*. tr = trace (< 0.1%).

Compound	%	Compound	%
α-Thujene	tr	trans-β-Ocimene	tr
α-Pinene	6.3	p-Cymene	0.8
Camphene	6.9	Terpinolene	0.3
β-Pinene	5.9	α-Thujone	tr
Sabinene	tr	β-Thujone	tr
Myrcene	4.1	Camphor	13.0
α-Terpinene	tr	Bornyl acetate	1.1
Limonene	2.9	β-Caryphyllene	3.0
1.8-Cineole	46.9	α-Humulene	1.0
cis-β-Ocimene	tr	Borneol + Terpinyl acetate	5.4
γ-Terpinene	1.2		
			98.8

The essential oil of a spontaneous hybrid of *S. fruticosa* and *S. officinalis* was found to have a unique composition (17): 1.8-cineole (20%), camphor (19%), α- and β-thujone (16%) were the main components. The oil obtained by steam distillation from freshly cut leaves of *S. fruticosa* was a colourless liquid with a harsh, spike rosemary, camphor-like odour (18). Buil et al. (14) found that *S. fruticosa* oil from Turkey was rich in 1.8-cineole (42%), camphor (9%), β-caryophyllene (8%) and β-pinene (5.5%); it grows wild in Sicily, Greece, Turkey, Lebanon and Algeria. Dalferth (19) reported that the major components of the essential oil of Greek origin were 1.8-cineole (64%), camphor (8%), β-pinene (8%) and α- and β-thujone (5%). The essential oil from commercial leaves contain 27.6-54.1% 1.8-cineole (20). This compound was also found in high concentration in some North American *Salvia* species (21).

Salvia dominica (= *S. graveolens*) is a strong smelling plant growing wild in chalky hills; the inflorescences and the leaves are rich in essential oil. The odour and composition of the oil resemble the oil distilled from the inflorescences of *S. sclarea* (Table III). The monoterpene hydrocarbons are found in low concentration in the oil (6.5%). The oil is rich in monoterpene alcohols and their acetates.

Table III. Percentages of the components of the essential oils from *Salvia dominica* and *S. sclarea*. tr = trace (< 0.1%).

Compound	*S. dominica*	*S. sclarea*	Compound	*S. dominica*	*S. sclarea*
α-Pinene	tr	tr	cis-Linalool oxide	tr	2.0
Camphene	tr	tr	trans-Linalool oxide	tr	1.3
β-Pinene	tr	tr	Linalool	25.9	19.1
Myrcene	2.6	tr	Linalyl acetate	28.4	60.2
α-Phellandrene	tr	tr	Terpinen-4-ol	1.1	tr
α-Terpinene	tr	tr	α-Terpineol	12.4	4.6
Limonene	0.8	tr	α-Terpinyl acetate	2.8	tr
1.8-Cineole	1.1	1.3	Neryl acetate	6.5	2.1
γ-Terpinene	2.2	1.2	Geranyl acetate	7.1	2.6
p-Cymene	tr	tr	Nerol	2.0	tr
Terpinolene	0.7	-	Geraniol	5.1	tr
				98.7	94.4

The characteristic aroma of the oil of *S. sclarea* is due to micro-components other than linalyl acetate and linalool (22); the same holds true for the oil of *S. dominica*. Although *S. dominica* oil is not known in commerce, it has a great potential as a substitute or additive to *S. sclarea* commercial oil.

The genus *Micromeria* is common in the Mediterranean basin. *M. fruticosa* grows wild on chalky rocks and flowers from February to June. Tea prepared from the fresh leaves has a minty aroma and is used in folk medicine for indigestion, coughs and colds and for lowering high blood pressure (3, 6). The main components in the oil distilled from fresh plants were pulegone and menthol (Table IV). The proportion of the monoterpene hydrocarbons was 14.4%. Pulegone, the minty component of *M. fruticosa*, is the industrial starting material of menthone and menthol (18).

Table IV. Composition of the essential oil from *Micromeria fruticosa*. tr = trace (< 0.1%).

Compound	%	Compound	%
α-Pinene	1.3	1-Octen-3-ol	0.4
Camphene	tr	Menthone	tr
β-Pinene	3.2	Isomenthone	3.7
Myrcene	1.0	Terpinen-4-ol	8.2
Limonene	8.0	β-Caryophyllene	8.8
cis-β-Ocimene	0.4	Menthol	10.1
γ-Terpinene	tr	Pulegone	45.9
p-Cymene	0.5	α-Terpineol	tr
Terpinolene	tr	Piperitone	2.8
3-Octanol	tr		
			94.3

REFERENCES

(1) Heath HB. 1978, Flavour technology: Profiles, Products, Application, Westport, Connecticut, Avi Publ. Co. Inc.
(2) Werker E, Ravid U, Putievsky E. 1985, Israel J. Bot., in press.
(3) Yaniv Z, Dafni A, Palevitch D. 1982, in: Aromatic Plants, Basic and Applied Aspects (Margaris N, Koedam A, Vokou D, Eds.), The Hague, Martinus Nijhoff Publ., p. 265.
(4) Ravid U, Putievsky E. 1983, Planta Med. 49:248.
(5) Putievsky E, Dafni A. 1979, The Book of Spices, Ramat-Gan, Israel, Massada Publ. (in Hebrew).
(6) Dafni A, Yaniv Z, Palevitch D. 1984, J. Ethnopharmacol. 10:295.
(7) Feinbrun-Dothan N. 1978, Flora Palestina, Part III, Jerusalem, The Israel Academy of Science and Humanities.
(8) Poulose AJ, Croteau R. 1978, Arch. Biochem. Biophys. 187:307.
(9) Husain SZ, Heywood VH, Markham KR. 1982, in:Aromatic Plants, Basic and Applied Aspects (Margaris N, Koedam A, Vokou D, Eds.), The Hague, Martinus Nijhoff Publ., p.141.
(10) Brieskorn CH, Brunner H. 1967, Planta Med. Suppl. 96.
(11) Zaitschik DV, Levontin S. 1971, Harokeach Haivri, 14:284 (in Hebrew).
(12) Fleischer A, Sneer N, Fleischer Z, Joffe A. 1980, Proc. 8th Internat. Congress Essential Oils, Cannes, France.
(13) Şarer E, Scheffer JJC, Baerheim Svendsen A. 1982, Planta Med. 46:236.
(14) Buil P, Garnero J, Guichard G, Konur Z. 1977, Riv. Ital. Essenze Profumi Piante Off. Aromat. Syndets Saponi Cosmet. Aerosols 59:379.
(15) Calzolari C, Staucher B, Pertioldi Marletta G. 1968, Analyst 93:311.

(16) Skrubis BC. 1972, Flavour Ind. 3:566.
(17) Putievsky E, Ravid U. 1984, Internat. Symp. on Conservation of Genetic Resources of Aromatic and Medicinal Plants, Oeiras, Portugal.
(18) Furia TE, Bellanca N, Eds. 1975, Fenaroli's Handbook of Flavour Ingredients, 2nd Ed., Cleveland, Ohio, CRC Press.
(19) Dalferth S. 1963. Ph.D. Thesis, University of Würzburg, FRG.
(20) Rhyu HY. 1979, J. Food Sci. 44:758.
(21) Lawrence BM. 1981, Essential Oils 1979-1980, Wheaton, Illinois, Allured Publ. Co.
(22) Formácek V, Kubeczka K-H. 1982, Essential Oil Analysis by Capillary GC and Carbon-13 NMR Spectroscopy, Chichester/New York/Brisbane/Toronto/Singapore, John Wiley & Sons.

STUDIES ON THE COMPOSITION OF ESSENTIAL OILS OF 'CINEOLIC EUCALYPTUS'

M.M. CARMO, S. FRAZÃO

ABSTRACT

In order to obtain information about the composition of 'cineolic eucalyptus' oils from different provenience, eight essential oil samples isolated by hydrodistillation were investigated. The results obtained by gas chromatography showed typical finger-prints although the content of the main constituent, 1.8-cineole, was similar. Characteristic data of the samples studied are given.

INTRODUCTION

In Portugal two species of the genus *Eucalyptus* are of great commercial importance. The plantations of *E. globulus* are about 30 000 ha and those of *E. maideni* about 7 500 ha. The essential oils obtained from the two species have a high content of 1.8-cineole and they are classified as 'cineolic eucalyptus' oils. The characteristics of these oils and their chromatographic finger-prints have been investigated in order to point out the differences in their chemical composition (1). This problem was first dealt with when the ISO standard 770 for *E. globulus* oil was worked out, and afterwards when the characteristics of rectified Portuguese eucalyptus oils were established (2, 3). A large number of analyses were carried out to set up the limits for the national and the ISO standards. More recently we studied also other commercially important 'cineolic eucalyptus' oils, obtained from different *Eucalyptus* species (4) namely *E. fruticetorum* from Australia and *E. smithii* from South Africa. We have also analyzed an oil sample from China, but its botanical origin was unknown. We are not sure if Spanish *E. globulus* oils are distilled from only this species; the values of the characteristics of these oils are less constant than those of the Portuguese *E. globulus* oils (5).

MATERIALS AND METHODS

The essential oil from Portuguese *E. globulus* was distilled in our laboratory and compared with the industrially available oil from the same

A. Baerheim Svendsen and J.J.C. Scheffer (eds), Essential oils and aromatic plants, ISBN-13: 978-94-009-5137-2

species. The samples from Australia and South Africa were obtained from Mr. Lasak, Biological and Chemical Research Institute, New South Wales Government, Australia. The sample from China and the Spanish oils were obtained from Dr. Peyron and Dr. Garcia Vallejo.

GC analyses were carried out on a gas chromatograph Perkin-Elmer Sigma 2B equipped with flame ionization detector and a packed column (2m). Conditions: column packed with 15% Carbowax 20M on Chromosorb W (80-100 mesh); oven temperature, $100^{\circ}C$; injector, $150^{\circ}C$; detector, $150^{\circ}C$; carrier gas, Nitrogen. For capillary GC of the Portuguese *E. globulus* oil a gas chromatograph Hewlett-Packard 5880 equipped with flame ionization detector was also used. Conditions: column (50 m) coated with Carbowax 20M; oven temperature, isothermal $60^{\circ}C$ (5 min), then 60-$200^{\circ}C$ ($2^{\circ}C$/min) and isothermal 10 min; injector, $220^{\circ}C$; detector, $200^{\circ}C$; carrier gas, Nitrogen.

RESULTS AND DISCUSSION

All samples studied were compared with the specifications given in the standard ISO 770 (essential oil of *E. globulus*) and the standard ISO/DIS 4732 (rectified Portuguese *E. globulus* oils). The specifications are listed in Table I. The Portuguese crude and rectified oils fully satisfied the specifications, the Spanish oils not always since they probably contain a high percentage of *E. maideni* oil (6, 7).

The characteristics of the other oil samples investigated are given in Table II. The chemical composition of these samples is summarized in Table III; the data were obtained by GC using the packed column.

Crude *E. globulus* oil differs from the corresponding rectified oils in containing sesquiterpenes. By capillary GC of this oil sample the following components were identified: α-pinene; limonene; 1.8-cineole; p-cymene; linalool; β-caryophyllene; aromadendrene; pinocarveol; alloaromadendrene; α-terpineol; globulol.

The *E. fruticetorum* oil had a typical odour and it was dextrorotatory. This is characteristic for the Australian oil and distinguishes it from the oil isolated from the same species growing in Spain (8). The sample satisfied the ISO standard for the Australian eucalyptus oil with a 1.8-cineole content of 80-85%.

The *E. smithii* oil has a peculiar odour and its characteristics are not standardized.

Table I. ISO standards for 'cineolic eucalyptus' oils.

ISO standard	Eucalyptus oils	Type of oil	Organoleptic characteristics	d_{20}^{20}	n_D^{20}	$[\alpha]_D^{20}$	Miscibility in ethanol 70%	Percentage of 1.8-cineole
770	*E. globulus*	crude	colourless to pale-yellow	0.906 0.925	1.4590 1.4670	0^0 $+10^0$	< 5	≥ 70
DIS 4732	*E. globulus*	rectified 70-75% cineole	colourless, clear, mobile liquid	0.907 0.917	1.4600 1.4640	$+3^0$ $+8^0$	< 8	70 75
DIS 4732	*E. globulus*	rectified 75-80% cineole	colourless, clear, mobile liquid	0.910 0.919	1.4590 1.4630	$+2^0$ $+6^0$	< 5	75 80
DIS 4732	*E. globulus*	rectified 80-85% cineole	colourless, clear, mobile liquid	0.913 0.920	1.4590 1.4620	$+2^0$ $+4^0$	< 3	80 85
3065	Australian	80-85% cineole	colourless to pale-yellow, clear, mobile liquid	0.918 0.928	1.4580 1.4650	-2^0 $+2^0$	< 3	80 85

Table II. Characteristics of different 'cineolic eucalyptus' oils. * = o-cresol method; ** = GC.

Eucalyptus species	Origin	Organoleptic characteristics	d_{20}^{20}	n_D^{20}	$[\alpha]_D^{20}$	Miscibility in ethanol 70%	Percentage of 1.8-cineole	
E. maideni	Portugal	pale-yellow, mobile liquid	0.917	1.4650	$+3^0$	3	79.0*	80.7**
E. fruticetorum	Australia	yellow, mobile liquid	0.918	1.4620	$+1^0$	3	81.0*	83.2**
E. smithii	South Africa	brownish-yellow, mobile liquid	0.905	1.4600	$+4^0$	> 20	78.0*	77.8**
unknown	China	colourless, mobile liquid	0.912	1.4580	-5^0	2.5	88.5*	87.4**

Table III. Percentage composition of different 'cineolic eucalyptus' oils.

Compound	*Eucalyptus* *maideni*	*fruticetorum*	*smithii*	'Chinese' sample
Isovaleric aldehyde	0.05	0.05	0.13	–
α-Thujene	0.03	0.03	0.14	–
α-Pinene	5.39	3.30	6.53	0.24
Camphene	0.33	0.15	0.05	0.09
β-Pinene	0.26	0.62	0.60	4.48
Myrcene	–	0.56	0.37	0.17
Limonene	4.90	4.03	6.40	1.72
1.8-Cineole	80.69	83.16	77.75	87.37
p-Cymene	1.90	5.44	2.25	5.25
Terpinen-4-ol	0.20	0.15	0.10	0.34
Pinocarveol Alloaromadendrene	3.52	1.30	0.67	–
α-Terpineol	2.14	1.20	2.91	–

CONCLUSIONS

The results obtained by GC and by the other methods confirmed that the ISO 770 and ISO/DIS 4732 standards are applicable only to the *E. globulus* oil. The Chinese oil sample investigated did not meet the ISO standards as to optical rotation. The Spanish crude eucalyptus oil has sometimes a low content of 1.8-cineole (68-69%) and the corresponding rectified oils also fail the limits slightly.

The *E. maideni* oil is most similar to the *E. globulus* oil but its GC finger-print is different. The *E. smithii* and *E. fruticetorum* oils have typical chemical compositions and odours. The ISO standards are not suitable for the Chinese and *E. smithii* oils.

It may be concluded that Portuguese *E. globulus* oil is the most uniform 'cineolic eucalyptus' oil as to chemical composition and physical characteristics.

REFERENCES

(1) Carmo MM. 1972, Anais da Acad. Brasileira de Ciências 44Suppl.:278.
(2) Proença da Cunha A. 1972, Bol. da Fac. Farm. Coimbra 32:5.
(3) Frazão S. 1981, Thesis, University of Montpellier, France.
(4) Frazão S. 1983, Thesis of 3rd cycle, University of Montpellier, France.
(5) Camino AY. 1972, Ion 370, 32:273.
(6) Frazão S, Carmo MM. 1982, in: 2ème Rencontre Intern. Pharm. de Montpellier, France.
(7) Garcia Martin D, Garcia Vallejo MC. 1982, in: Annales techniques 8. Congrès international des huiles essentielles, Cannes 1980, p. 377.
(8) Garcia Vallejo MC, Blanco Diez MM, Garcia Martin D. 1982, in: Annales techniques 8. Congrès international des huiles essentielles, Cannes 1980, p. 271.

BIOLOGY AND ESSENTIAL OIL OF *HUMULUS LUPULUS* VAR. *NEOMEXICANUS*

K. KNOBLOCH, H. PAULINI, C. ELEY, J. ELEY, E. ZIEGLER, H. BRANDAUER,
K. MICHAELIS, O. VOSTROWSKY

ABSTRACT

Humulus lupulus var. *neomexicanus* grows in the semi-arid climate of Central Colorado. Its production of essential oil shows a characteristic seasonal dependence. This has been observed over two vegetation periods when the essential oil from the strobiles was isolated by steam distillation. The main components of the essential oil were identified by GC and GC-MS. More than 40 different components were detected, including monoterpene hydrocarbons and alcohols, ketones, methyl esters of saturated and unsaturated carboxylic acids; in addition, esters of short chain acids with monoterpene alcohols and sesquiterpene hydrocarbons were found. Although the typical and chief scent of the neomexican hops strobiles is that of mango (*Mangifera indica*) fruits, the main components of the essential oil resemble those of the hop varieties already analyzed.

'Wild Hops' or 'Wild Hop-Vine' is the term used for *Humulus lupulus* L. var. *neomexicanus* Nels. & Cockerell, which is the only native genus and species of the Cannabaceae family within the Central Rocky Mountains. It grows at an altitude of 1300-3000 m, and is found in canyons and foothills where it is seen climbing over bushes. The annual aerial climbing stems arize from the perennial rhizome and become 5-10 m long. The opposite leaves are lobed palmately. Two of the female flowers sit together under a large persistent bract. At maturity the bracts from the conelike 'hops' change colour and appear as papery and conspicuous clusters known as strobiles (1, 2).

The essential oil is produced by the female inflorescences, the strobiles, *strobuli lupuli*, (3) which contain the oil bearing glands, *glandulae lupuli* (4). For hundreds of years strobiles of *H. lupulus* L. have been used to flavour beer and to calm nervous disorders. The essential oil is known to possess narcotic as well as diuretc, antiseptic and bittertonic activity. The strobiles of *H. lupulus* var. *neomexicanus* reveal a differnt scent when compared with that of *H. lupulus*. Surprisingly, they smell like mango fruits (*Mangifera indica* L.). Until recently no-

A. Baerheim Svendsen and J.J.C. Scheffer (eds), Essential oils and aromatic plants, ISBN-13: 978-94-009-5137-2

thing was known on the composition of its essential oil. In a first contribution (5) we reported on the main volatile substances and analyzed them quantitatively; in addition, we observed a characteristic seasonal dependence of the essential oil production of the strobiles.

MATERIALS AND METHODS

Humulus lupulus L. var. *neomexicanus* Nels. & Cockerell was collected in the semi-arid climate of Western El Paso County of Central Colorado (5) from January 1980 through October 1981. The collection site was located at 2423 m elevation on a north facing slope. The soil consisted mainly of decomposed granite with sand as the major texture component.

The essential oils were isolated from 10-15 g of freeze-dried strobiles by steam distillation for 2 h. The essential oils were collected by dissolving in 0.20 ml xylene within a graduated capillary tube. For GC analysis and GC-MS, see (5).

RESULTS AND DISCUSSION

Samples taken in October (1980 and 1981) were already air-dried due to the arid climate of Central Colorado. The content of essential oil was about 1-2 µl per 1 g of air-dried samples collected during the winter season and stayed at this level until the end of July. In general, a seasonal dependence in the oil biosynthesis was obvious. The biosynthesis of the essential oil in female buds started in early August and reached a maximum of 6-8 µl per 1 g of dried material in late summer.

All samples collected were chromatographed on a capillary WG-11 column (comparable to FFAP) and on a packed OV-17 column. Thereby the main components of the oil were determined quantitatively. The oil from a sample taken on August 27, 1980 was chosen for a complete analysis of its different components by GC-MS (Table I).

For identification of the main components the mass spectra were compared with those already reported in the literature (6-8) and with those obtained with authentic samples (9). In addition, the elution sequence of each compound was examined by comparison with data from the literature obtained on columns of similar polarity (10).

Myrcene (62.18%) was identified as the main substance, other monoterpene hydrocarbons were α-pinene, sabinene, limonene and E-ocimene (total amount about 3%). Linalool (1.09%) and geraniol (1.52%) were the only monoterpene alcohols detected.

Table I. Composition of the essential oil of *H. lupulus* var. *neomexicanus* harvested August 27, 1980. Conditions: glass capillary column coated with SE-30; temperature program, 50-220°C (1.5°C/min) (5), a = spectrum of authentic sample; b = spectrum in literature (6-8); c = retention indices on OV-101 (10).

Compound	Percentage	Identification method	Retention index
C_9H_{12}	0.53	GC-MS	
$C_7H_{16}O_2$	0.33	GC-MS	
α-Pinene	0.26	GC-MS a,b	942 c
C_9H_{12}	0.21	GC-MS	
Sabinene	1.39	GC-MS a,b	976 c
Myrcene	62.18	GC-MS a,b	
Isobutyrate	0.38	GC-MS	
Methyl heptanoate	0.37	GC-MS b	1006 c
Limonene	0.54	GC-MS a,b	1030 c
E-Ocimene	0.95	GC-MS a,b	1038 c
Branched methyl octanoate	0.95	GC-MS	
Linalool	1.09	GC-MS a,b	1092 c
Methyl octanoate	0.24	GC-MS b	1107 c
Branched methyl nonanoate	0.14	GC-MS	
Methyl nonanoate	0.12	GC-MS b	1207 c
Ketone $C_{11}H_{23}O$	0.05	GC-MS	
Geraniol	1.52	GC-MS a,b	1243 c
Undecan-2-one	0.24	GC-MS b	1276 c
Methyl decenoate	1.10	GC-MS b	
Methyl decadienoate	1.65	GC-MS	
Methyl decadienoate	0.22	GC-MS	
Methyl decanoate	0.31	GC-MS b	1307 c
Acetate of geraniol or nerol	0.24	GC-MS a,b	
α-Ylangene	0.13	GC-MS a,b	
α-Copaene	0.52	GC-MS a,b	1398 c
β-Caryophyllene	2.90	GC-MS a,b	1428 c
β-Cubebene (?)	0.26	GC-MS b	
Humulene	2.12	GC-MS a,b	1465 c
Propionate of geraniol or nerol	0.36	GC-MS	
γ-Muurolene (?)	1.21	GC-MS b	1475 c
$C_{15}H_{24}$	0.87	GC-MS	
$C_{15}H_{24}$	0.98	GC-MS	
Isobutyrate of geraniol or nerol	0.51	GC-MS	
γ-Cadinene and Calamene	2.36	GC-MS a,b	1518 c
δ-Cadinene	1.43	GC-MS a,b	1524 c
$C_{15}H_{24}$	0.45	GC-MS	
Unknown	0.92		
Unknown	0.70		

The essential oil of *H. lupulus* var. *neomexicanus* appears to be similar to other varieties of hops as far as its monoterpene composition is concerned (11-16). Likewise, the so-called typical ingredients of the essential oil from hops could be identified in the variety *neomexicanus*. These are the methyl esters of heptanoic, octanoic, nonanoic and decanoic acid, both the methylbranched alkanoic methyl esters, the methyl esters of the unsaturated carboxylic acids, as well as the acetate, propanoate and isobutyrate of geraniol and nerol.

The sesquiterpene fraction of the oil was relatively complex. We were able to identify α-ylangene, α-copaene, β-caryophyllene, humulene, γ- and δ-cadinene by means of comparison with authentic samples. β-Cubebene, γ-muurolene and calamene could be assigned only by comparing the mass spectra with published data (6-8) or known retention indices (10).

American *H. lupulus* cultivars appear to be of hybrid origin. Hybridization occurred apparently between European cultivars and one of the three major indigenous varieties of wild North-American plants; these are the varieties *neomexicanus*, *pubescens* and *lupuloides*(17).

Studies on five groups of wild hops and on three groups of cultivars revealed (17):

- Japanese cultivars are associated uniformly with the indigenous Japanese *H. lupulus* var. *cordifolius*;
- European cultivars are associated consistently with the European *H. lupulus* var. *lupulus*;
- North-American cultivars joined consistently the North-American *H. lupulus* var. *lupuloides*, and in addition, North-American *H. lupulus* var. *neomexicanus* appeared to be associated with this pair (Fig. 1).

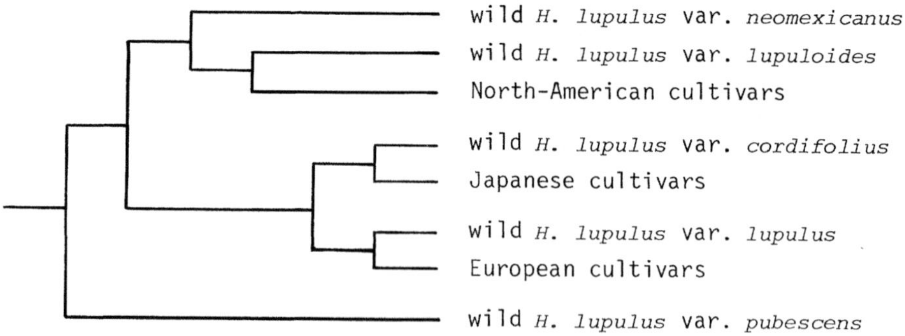

Fig. 1. Relationships between wild samples of five taxa of *Humulus lupulus* and geographical groups of cultivars (17).

Summarizing we can state that in spite of its very special scent, the variety *neomexicanus* does not appear distinctly different as far as the chemical composition of the essential oil is concerned. More sensitive analyses may help to characterize the interesting hop variety *neomexicanus*.

ACKNOWLEDGEMENT

This work was supported partly by the 'Adalbert-Raps-Stiftung', Kulmbach, FRG.

REFERENCES

(1) Harrington HD. 1964, Manual of the Plants of Colorado, Chicago, The Swallow Press Inc.
(2) Weber WA. 1976, Rocky Mountain Flora, Boulder, Colorado, Colorado Associated University Press.
(3) Wagner H. 1980, Pharmazeutische Biologie, Vol. 2 (Drogen und ihre Inhaltstoffe, Stuttgart - New York, Gustav Fischer Verlag.
(4) Teuscher E. 1979, Pharmazeutische Biologie (Borris H. Ed.), Braunschweig - Wiesbaden, Friedr. Vieweg & Sohn.
(5) Knobloch K, Paulini H, Eley C, Eley JH, Ziegler E, Brandauer H, Michaelis K, Vostrowsky O. 1982, Z. Naturforsch. 37c:565.
(6) Thomas AF, Willhalm B. 1964, Helv. Chim. Acta. 47:475.
(7) Stenhagen E, Abrahamsson S, McLafferty FW. 1974, Registry of Mass Spectral Data, New York, J. Wiley & Sons.
(8) Cornu A, Massot R. 1966, Compilation of Mass Spectral Data, London, Heyden & Sons Ltd.
(9) Vostrowsky O, Michaelis K, Ihm H, Zintl R, Knobloch K. 1981. Z. Lebensm.-Unters. Forsch. 173:365.
(10) Jennings W, Shibamoto T. 1980, Qualitative Analysis of Flavor and Fragrance Volatiles by Glass Capillary Chromatography, New York, Academic Press.
(11) Stevens R. 1967, Chem. Rev. 67:19.
(12) Buttery RG, McFadden WH, Teranishi DR, Kealy MP. 1963. Nature 200: 435.
(13) Masada Y. 1967, Analysis of essential oils by gas chromatography and mass spectrometry, New York, John Wiley & Sons, p. 226.
(14) Howard GA, Tachell AR. 1956, J. Inst. Brewing. 62:158.
(15) Sharpe F, Laws DRJ. 1981, J. Inst. Brewing. 87:96.
(16) Goedkoop W. 1960, Tijdschr. Brouw. en Mout 20:39
(17) Small E. 1980, Can. J. Bot. 58:676.

SESQUITERPENES AND OTHER VOLATILE METABOLITES FROM LIQUID CULTURES OF *CERATOCYSTIS POPULINA* (ASCOMYCOTA) - ESSENTIAL OIL COMPOUNDS FROM FUNGI

H.-P. HANSSEN

ABSTRACT

The ascomycete *Ceratocystis populina* was grown on a defined synthetic liquid culture medium. Volatile metabolites were obtained by steam distillation and identified or characterized by GLC and MS data in comparison with authentic substances. Major components were sesquiterpenes with a 1,7-dimethyl-4-isopropyldecaline skeleton (T-muurolol, α-amorphene and probably γ-amorphene).

Several *Ceratocystis* (Ascomycota; Ophiostomales) species are known to produce volatile metabolites responsible for the characteristic odour of the cultures. In the past numerous constituents have been identified including short-chain alcohols and esters ('fruit esters'), acyclic mono- and sesquiterpenes, terpenoids (methyl heptenyl compounds), and 2-phenylethyl acetate (1).

The species *C. populina* was isolated for the first time by Hinds and Davidson from aspen (2). Strain RWD 835 B produces a pleasant, fruit-like fragrance when cultivated on a defined synthetic culture medium (3). The identification of the steam volatile constituents by GC and GC-MS revealed a composition of volatiles differing totally from all other *Ceratocystis* strains (nearly 40 belonging to 10 species) hitherto investigated.

MATERIAL AND METHODS

Ceratocystis populina RWD 835 B was isolated from aspen in Colorado, USA by Davidson.

The ascomycete was grown on a defined synthetic liquid culture medium containing glucose (2%), asparagine (0.1%), thiamine and mineral salts (3) for 24 days. The volatile metabolites were determined after 4, 10, 17 and 24 culture days. In each case, they were obtained from 20 cultures (grown in 250 ml erlenmeyerflasks containing 50 ml culture broth) by circulation steam distillation (4) in 2 ml pentane.

A. Baerheim Svendsen and J.J.C. Scheffer (eds), Essential oils and aromatic plants, ISBN-13: 978-94-009-5137-2

After separation into five fractions of different polarity by dry-column chromatography (5), the steam distillates were analyzed by GLC. The identification of the various components resulted from GLC and GC-MS data compared with those of authentic substances or the literature (6-8).

GLC analyses were performed using a Perkin-Elmer F 22 apparatus equipped with a glass capillary WG-11 column (22 m x 0.3 mm i.d.), a flame ionization detector and a computing integrator (Autolab System I, Spectra Physics). Operating conditions: linear temperature program 80-200°C, 2°C/min or isothermal; injector, 180°C; detector, 180°C; carrier gas, Nitrogen, 1 ml/min; injection volume, 1.0 µl.

MS analyses were carried out on a Varian MAT 111 (GNOM) mass spectrometer (80 eV) using a 3 m packed Carbowax 20M (3%) column under various isothermal conditions or a temperature program as described above.

Quantities of volatile constituents were calculated via an internal standard (6-methyl-5-hepten-2-one) using FID-specific substance factors.

RESULTS AND DISCUSSION

The mycelium weights and the total amounts of volatiles were determined after 4, 10, 17 and 24 days. In Fig. 1 the data obtained are given. The steam distillates consisted of about 30 components. Fig. 2 shows a gas chromatogram of the concentrated sample obtained from 17 days old cultures.

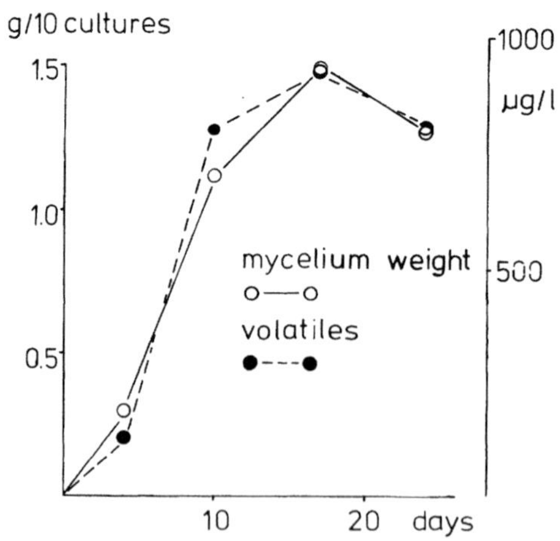

Fig. 1. Mycelium weights and amounts of volatiles in cultures of *C. populina* after 7, 10, 17 and 24 days.

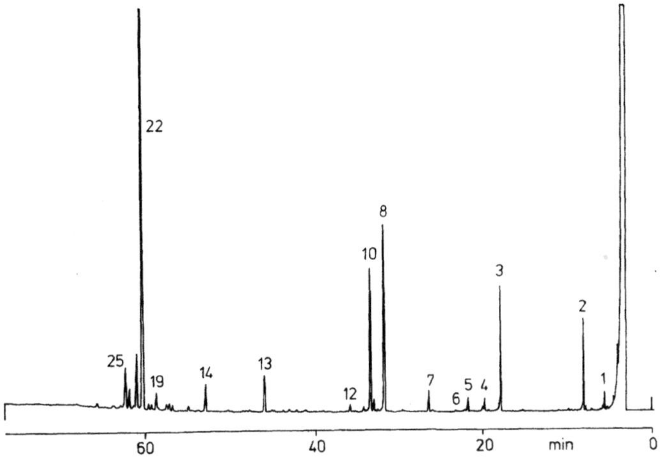

Fig. 2. Chromatogram of the volatiles in a 17 days old culture of *C. populina*. For gas chromatographic conditions, see under Material and Methods.

Predominating constituent is the bicyclic sesquiterpene alcohol T-muurolol (peak 22) that could be identified already from cultures of the brown rot fungus *Lentinus lepidus* (9). Table I indicates that this compound amounted to approximately 430 µg/l culture medium representing about 46% of the total yield of volatiles. Other peaks could be characterized as sesquiterpene alcohols as well, one of them tentatively as δ-cadinol (peak 23).

Table I. Composition of steam distillates obtained from 17 days old cultures of *C. populina* RWD 835 B grown on a glucose-asparagine-mineral salt medium; tr = trace.

Peak No.	Compound	µg/l	%
1	Isobutanol (?)	18.5	2.0
2	Isoamyl alcohol	54.7	5.9
3	1-Octen-3-ol	34.6	3.7
6	Linalool	tr	
7	C_5H_7NO	14.9	1.6
8	α-Amorphene	89.8	9.6
10	γ-Amorphene (?)	68.0	7.3
12	δ-Cadinene	tr	
13	$C_{15}H_{20}O$	23.8	2.4
22	T-Muurolol	427.4	45.8
23	δ-Cadinol (?)	123.2	13.2
25	$C_{15}H_{22}O$	27.7	3.0
	Minor compounds		5.5

Other sesquiterpenes were found in the hydrocarbon fraction, the main components being α-amorphene (peak 8) and probably γ-amorphene (peak 10), and δ-cadinene in traces. Linalool (peak 6) was another terpene constituent.

Other volatiles, identified or chemically characterized, were isoamyl alcohol (peak 2) and 1-octen-3-ol (peak 3), an ubiquitous volatile constituent of fungi. An unknown nitrogen-containing compound (peak 7) is probably a pyrrol derivative.

The composition of the volatiles accumulated by *C. populina* RWS 835 B differs from all *Ceratocystis* strains hitherto investigated, especially from those with Chalara anamorphs (*Ceratocystis s.s.*) (1). Further investigations should prove whether species having a Chalara conidial state can be distinguished from others of this genus by their volatiles. After all, morphological and biochemical data have supported a partition of the genus *Ceratocystis* (10, 11).

So far, volatile sesquiterpenes with an amorphane skeleton have never been reported to be present in fungi. In higher plants, these 'antipodic' compounds are relatively seldom compared with the related sesquiterpenes with a cadinane structure (12). The yields of volatiles obtained from strain *C. populina* RWD 835 B were relatively low under the culture conditions chosen (ppb-range). As shown for other species, they could possibly be enhanced by selecting suitable strains and/or optimizing the culture conditions (13, 14).

ACKNOWLEDGEMENTS

We are thankful to Dr. W. Bruhn, Dragoco, Holzminden, FRG for authentic substances and for valuable discussion. This work was supported by the 'Deutsche Forschungsgemeinschaft'.

REFERENCES

(1) Sprecher E, Hanssen H-P. 1983, Antonie van Leeuwenhoek 49:493.
(2) Hinds TE, Davidson RW. 1967, Mycologia 59:1102.
(3) Sprecher E. 1959, Planta 53:565.
(4) Sprecher E. 1963, Dtsch. Apoth.-Ztg. 103:213.
(5) Kubeczka K-H. 1973, Chromatographia 6:106.
(6) Stenhagen E, Abrahamsson S, McLafferty FW. 1976, Registry of Mass Spectra Data, New York, J. Wiley & Sons.
(7) Noever de Brauw MC ten, Bouwman J, Tas AC, La Vos GF. 1979-1982, Compilation of Volatile Compounds in Food, Vol. I-XI, Central Institute for Nutrition and Food Research - TNO, Zeist, The Netherlands.
(8) Borg-Karlson A.-K, Norin T, Talvitie A. 1981, Tetrahedron 37:425.

(9) Hanssen H-P, unpublished.
(10) Weijman ACM, Hoog GS de 1975, Antonie van Leeuwenhoek 41:353.
(11) Hoog GS de, Scheffer RJ. 1984, Mycologia 76:292.
(12) Klein E, Schmidt W. 1973, Dragoco Rep. 20:3.
(13) Hanssen H-P., Sprecher E. 1981, in: Flavour '81 (Schreier P, Ed.), Berlin/New York, Walter de Gruyter, p. 547.
(14) Sprecher E, Hanssen H-P. 1982, Planta Med. 44:41.

EFFECT OF DRYING AND FREEZE-DRYING ON THE AROMA OF DILL – *ANETHUM GRAVEOLENS* CV MAMMUT

R. HUOPALAHTI, E. KESÄLAHTI

ABSTRACT

Aroma compounds of fresh, freeze-dried and dried dill (*Anethum graveolens* L. cv Mammut) were isolated by solvent extraction with a mixture of n-pentane and diethyl ether followed by concentration and a clean-up procedure. Combined fused silica capillary gas chromatography-mass spectrometry was employed for identification of the aroma compounds. 70 volatile compounds were detected and 13 of them were identified: α-pinene, α-phellandrene, limonene, β-phellandrene, 3,6-dimethyl-2,3,3a,4,5,7a-hexahydrobenzofuran and myristicin as the major components. During both drying processes some not yet identified compounds increased in amounts. The almost total loss of the benzofuran derivative, one of the most abundant and important compounds in dill, was the largest change in the composition of the volatile constituents. The absolute amount of aroma compounds in freeze-dried dill was four times larger than in dried dill, while that in fresh dill was ten times larger.

INTRODUCTION

Dill, *Anethum graveolens* L., is in Finland a popular aromatic herb that is largely used for flavouring of various food products such as sauces, soups, pickled vegetables and sea foods. Usually the essential dill oil, distilled from fresh whole herb or ripe fruits, is preferred by the food industry. In Finland, however, domestic dill is mainly consumed as fresh whole plant before the bud formation or at the flowering stage (1) during a short harvesting period, while small amounts of chopped dill leaves, dried or frozen, are also used.

Essential oils of herbs mainly consisting of terpenoids are unstable on exposure to heat and light; so, the aroma of dill markedly decreases or changes during processing such as drying and freezing.

The main constituents of fresh dill before flowering are α-phellandrene and 3,6-dimethyl-2,3,3a,4,5,7a-hexahydrobenzofuran (1-4). The preliminary studies on the effects of different processes on the aroma constituents of dill were performed by Drawert et al. (5) and Schreier et al. (6).

The present work was undertaken to study the changes occurring in the

A. Baerheim Svendsen and J.J.C. Scheffer (eds), Essential oils and aromatic plants, ISBN-13: 978-94-009-5137-2

composition and content of the volatile compounds in dill processed by drying with hot air and freeze-drying, in order to find the optimum unit operations for an efficient preservation of a high quality of dill.

MATERIALS AND METHODS

Dill (*Anethum graveolens* L. cv Mammut) was cultivated in Central Finland (Sahalahti, 61°30' N.lat.) on a prefertilized light sphagnum peat, huminocity 1-3 (Finn Peat, ST-400-82). The whole plants were harvested before flowering and either air-dried or freeze-dried. The drying conditions are presented in Table I. The dried dill samples were packed in Aluminium foil laminate (7) and stored at room temperature. Analyses were carried out during the week after the harvest.

Aroma compounds were isolated by extracting 45 g of fresh dill or 10 g of dried dill for 6 h with 350 ml of a mixture of redistilled n-pentane and diethyl ether (1:2 v/v) by using a modified Soxhlet-technique in which the side tube of the regular apparatus was replaced by a column containing glass beads to prevent the cycle of aroma compounds. The extract was concentrated at 35°C with a Widmer column to 2 ml. A purification of the concentrate was carried out at 10°C on a silica gel column as described by Scheffer et al. (8). Detailed information is available in our previous papers (1, 2).

GLC analyses of the aroma concentrates were carried out on a gas chromatograph Varian 3700 equipped with flame ionization detector and connected to an integrator Hewlett-Packard 3388A. Column: fused silica (25 m x 0.32 mm i.d.) coated with OV-351 (film thickness 0.20 μm); column temperature, 70°C (2 min) then 70-230°C (2°C/min); carrier gas, Nitrogen, 1.5 ml/min.

Mass spectra were recorded at 70 eV on a VG Analytical 7070E instrument using the same GC column.

The total aroma content was calculated by using linalool as an internal standard and summing up 70 corresponding peaks of each chromatogram. Identification methods have been discussed in more detail in a previous paper (1).

RESULTS AND DISCUSSION

The moisture content of the dill samples (fresh, air-dried and freeze-dried respectively) and the drying temperatures are presented in Table I, as well as the absolute amounts of the volatile compounds as compared with the absolute amount in fresh dill (= 100%), and the total amounts of the aroma compounds in the samples studied.

Table I. Characteristics of fresh, air-dried and freeze-dried dill.

Dill	Moisture content %	Drying temperature ^{o}C	Total amount of aroma compounds mg/kg dry weight	Relative amount of aroma compounds %
Fresh	89.7	-	3259	100.0
Air-dried	11.1	25	489	15.0
Air-dried	10.2	40	290	9.0
Air-dried	11.8	50	366	11.2
Freeze-dried I	2.2	-30	825	25.3
Freeze-dried II	16.0	-30	1879	57.6

The total amount of aroma compounds in air-dried dill samples varied from 290 mg/kg dry weight to 489 mg/kg dry weight, depending on the drying temperatures; so, the lowest aroma content was obtained when the drying temperature was 40°C, and the highest content when the drying temperature was 25°C. The total amount of aroma compounds was 825 mg/kg dry weight in freeze-dried sample I, where the moisture content was 2.2%. The corresponding value was 1879 mg/kg dry weight in sample II, where the moisture content was 16.0%. The optimum moisture content of dried herbs is about 10%, since the growth of different moulds is possible at higher percentages, and rehydration becomes more difficult at lower percentages.

Typical gas chromatograms of the aroma concentrates of fresh, freeze-dried and air-dried dill (chromatograms A, B and C respectively) are shown in Fig. 1. In each concentrate 70 compounds were detected between the solvent peak (S) and the point indicated by an arrow. 13 components were identified of which α-pinene, α-phellandrene, limonene, β-phellandrene, 3,6-dimethyl-2,3,3a,4,5,7a - hexahydrobenzofuran and myristicin were the main ones. During both drying processes also some not yet identified compounds increased in amounts (peaks 13, 14 and 15; only their molecular weights were determined). The amounts of these aroma compounds are presented in Table II. α-Phellandrene was the main compound in fresh, freeze-dried II and air-dried dill, whereas in freeze-dried sample I the most abundant compound was peak 13. The second abundant compound in fresh dill was the benzofuran derivative. This compound has been reported to exhibit an odour similar to fresh dill herb (1, 3) and it is therefore advisable to preserve this compound during dill processing. It is, however, evident from the data given in Table II, that severe losses of aroma substances occur during freeze-drying and air-drying. The amount of this important compound was 3.6% (freeze-dried I) and 22.5% (freeze-dried II) of the

Table II. Amounts of some aroma compounds (mg/kg dry weight) in fresh, freeze-dried and air-dried dill. Numbering of the compounds corresponds to Fig. 1; tr = trace (< 1.0 mg/kg dry weight); * = 3,6-dimethyl-2,3,3a,4,5,7a-hexahydrobenzofuran.

Peak No.	Compound	Fresh	Freeze-dried I	Freeze-dried II	Air-dried 25°C	Air-dried 40°C	Air-dried 50°C
1	α-Pinene	57.8	6.1	30.5	11.7	6.4	13.6
2	Terpinolene	3.9	5.4	3.1	4.4	3.7	2.4
3	β-Pinene	12.0	4.5	5.1	5.0	4.0	4.0
4	n-Undecane	5.7	tr	3.3	tr	tr	tr
5	α-Phellandrene	1980.7	149.1	416.0	133.0	60.8	80.6
6	Limonene	100.3	6.9	20.0	6.8	2.9	4.0
7	β-Phellandrene	275.3	17.6	65.1	21.6	9.0	11.2
8	γ-Terpinene	1.6	tr	tr	tr	tr	tr
9	p-Cymene	54.8	1.4	40.2	11.4	3.1	4.0
10	Benzofuran derivative*	397.9	14.4	89.4	4.9	tr	tr
11	α-Terpineol	12.2	2.2	5.7	1.9	1.3	tr
12	Carvone	tr	1.6	tr	tr	tr	tr
13	MW = 278	10.4	260.0	382.4	63.3	38.7	26.4
14	MW = 278	2.3	38.0	48.3	5.2	3.0	8.4
15	MW = 278	3.1	59.0	65.1	7.6	4.5	9.5
16	Myristicin	44.0	15.1	42.6	5.7	2.7	3.0

original value determined in fresh dill herb. In air-dried dill the corresponding values were 1.2% (25°C) or only traces (40°C and 50°C). Drawert et al. (5) found more severe losses: 0.8% in air-dried dill and 1.8% in freeze-dried dill; the temperature of the air was not reported.

According to the present study the losses of the main component in fresh dill, α-phellandrene, were also drastic: 7.5% (freeze-dried I), 21.0% (freeze-dried II), 6.7% (25°C), 3.1% (40°C) and 4.1% (50°C) of the fresh dill herb value were left. Losses of other dill aroma compounds were of the same order.

The increase of the amounts of the unidentified compounds 13, 14 and 15 was remarkable. Especially in the freeze-dried samples 13 was one of the most abundant components, and so it evidently has also influence on the aroma of freeze-dried dill. The identification of that compound is our next prime subject.

The results obtained show that the total aroma content of freeze-dried dill was about four-fold that of dried dill. In addition, the amount of 3,6-dimethyl-2,3,3a,4,5,7a-hexahydrobenzofuran, one of the most abundant and important aroma compounds in dill, was much higher in the freeze-dried samples than in the air-dried ones. So, freeze-drying is superior to air-drying in dill processing.

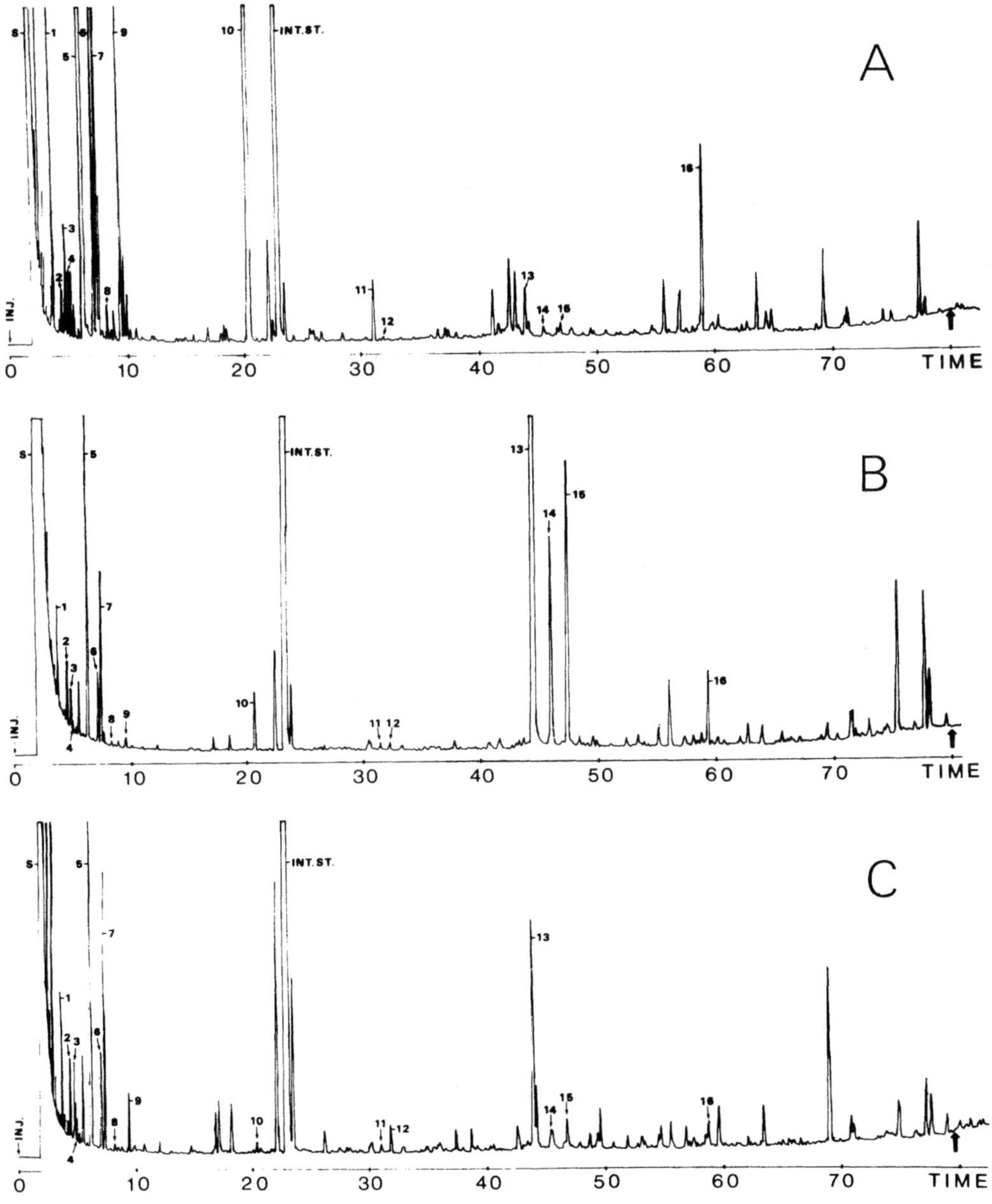

Fig. 1. Gas chromatograms of aroma concentrates isolated from fresh (A), freeze-dried (B) and air-dried (C) dill. Int. st. = internal standard (linalool); only the peaks between the solvent peak (S) and the point indicated by an arrow are included in the calculations.

ACKNOWLEDGEMENT

We thank Pentti Oksman for the GC-MS analyses.

REFERENCES

(1) Huopalahti R, Linko RR. 1983, J. Agric. Food Chem. 31:331.
(2) Huopalahti R, Kallio H, Kärppä P, Linko RR. 1981, in: Flavour '81 (Schreier P, Ed.), Berlin/New York, Walter de Gruyter, p. 369.
(3) Schreier P, Drawert F, Heindze I. 1981, Lebensm. Wiss. Technol. 14:150.
(4) Porter NG, Shaw ML, Shaw GJ, Ellingham PJ. 1983, New Zeal. J. Agric. Res. 26:119.
(5) Drawert F, Schreier P, Bhiwapurkar S, Heindze I. 1981, in: Flavour '81 (Schreier P, Ed.), Berlin/New York, Walter de Gruyter, p. 649.
(6) Schreier P, Drawert F, Heindze I. 1981, Z. Lebensm. Unters. Forsch. 172:257.
(7) Huopalahti R. 1985, in: Progress in Flavour Research 1984 (Adda J, Ed.), Amsterdam, Elsevier, p.309.
(8) Scheffer JJC, Koedam A, Schüsler MThIW, Baerheim Svendsen A. 1977, Chromatographia 10:669.

Note from the Editors: the benzofuran derivative mentioned in this paper has been named 'dill ether' by Brunke EJ and Rojahn W. 1984, Dragoco Rep. 31:67.

DIFFERENCES IN THE YIELD OF PLANT MATERIAL, ESSENTIAL OILS AND THEIR MAIN COMPONENTS DURING THE LIFE CYCLE OF *ORIGANUM VULGARE* L.

E. PUTIEVSKY, U. RAVID, S.Z. HUSAIN

ABSTRACT

Ten strains of *Origanum vulgare* L. selected from wild populations were studied for the yield of plant material, essential oils and their main components under intensive conditions during two years of growth. They were divided into three groups according to their characteristics. The characters examined - fresh weight, biomass ratio between various parts of the plant, yield of essential oil and their main constituents - show considerable variation among the three groups. The connections between the phenology, harvest date and yield, and percentage of essential oil components are discussed.

INTRODUCTION

The cultivation of *Origanum vulgare* L. in the Mediterranean region has been studied extensively in recent years (1). During summer and autumn there is no phenological evidence to indicate the appropriate time for harvesting. In these seasons plants are harvested when the oldest leaves start to fall (2). In the spring, 4-5 months after the previous harvest, the plants start to flower. In places where *Origanum* is collected from wild populations (Greece, Turkey), harvesting starts between the end of the flowering period and the beginning of the seed formation (3, 4).

The work presented here was carried out to determine the optimum date for spring harvesting in order to obtain the highest yield of shoots and essential oil. The data collected from the various strains represent the conditions in variable populations.

MATERIAL AND METHODS

Ten strains selected from wild populations of *Origanum vulgare* in Greece, were propagated vegetatively by stem cuttings in the winter 1977, in plots 2.5 m x 10 m (5). During 1977 and 1978 each strain was harvested 4 times; the last harvest was in November 1978. The fertilization, irrigation and harvesting procedures during 1977-1979, are described elsewhere (5).

A. Baerheim Svendsen and J.J.C. Scheffer (eds), Essential oils and aromatic plants, ISBN-13: 978-94-009-5137-2

The results presented in this paper refer to the third year of growth, i.e. 1979. From early spring (April) until the middle of the summer (July), 5 samples (0.5 m x 1.0 m) were taken every 4 week from each strain, and weighed. Leaves, stems and inflorescences were separated and weighed immediately after the harvest. Samples (4 per strain) were dried at 40°C for 48 h and the content of essential oil in the leaves was determined with a Clevenger-type apparatus. Samples of the essential oil were analysed by GLC (5).

The 10 strains were divided into 3 groups - group A: strains 4, 7, 8 and 10; group B: strains 2, 3, 5, 6, and 9; group C: strain 1 (5). The data were collected for each strain, but are presented according to the groups mentioned.

RESULTS AND DISCUSSION

The first harvest was made at the end of the winter (April), in which the fresh yield in the three groups reached its maximum level (Fig. 1). As the time passed, the fresh yield decreased. The fresh yield in group C was almost the double of that in group A and B, but the leaf-percentage was almost three times greater in group B (Fig. 2). It is clear that whereas the leaf ratio drops when the plants are not harvested, the inflorescence ratio increases. In group A and B it happened during May and June, but in group C it happened before April. The stem ratio showed no change and was very similar between the three groups (Fig. 2).

The essential oil content in two of the three groups reached its maximum level at the beginning of May, when the plants were flowering (Table I). The content of essential oil was highest in group A and lowest in group C. There were great differences in the phenols and to some extent in the content of monoterpene hydrocarbons between the three groups. Even so, there were no great changes in each of the main components during maturity. The content of phenols increased in the three groups while the content of monoterpene hydrocarbons decreased in group A and B, and increased in group C.

At the end of the winter, when it was possible for the first time to harvest the *Origanum* plants mechanically, they reached the maximum yield level of fresh as well as dried leaves. In some groups it was even possible to delay the harvest until May with its inhibited regrowth ability in cultivation conditions (2, 6). Collection of *Origanum* from natural

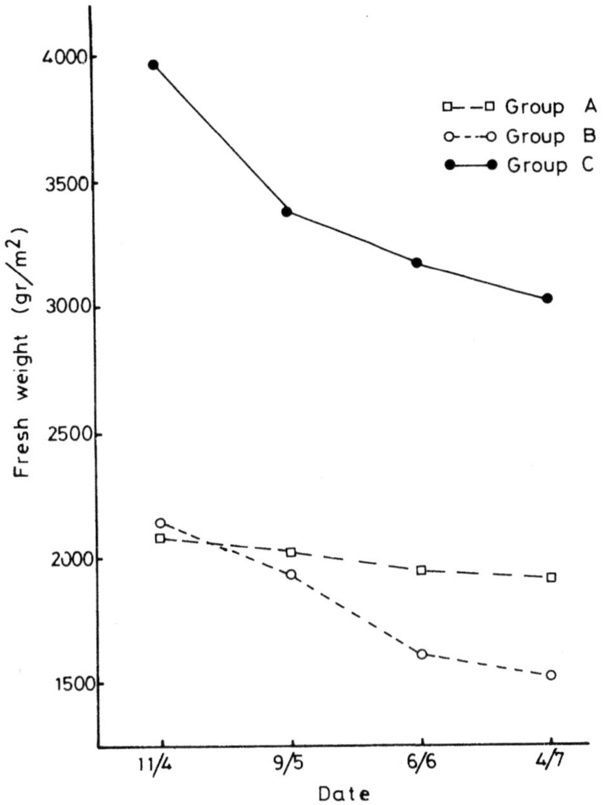

Fig. 1. Changes of fresh weight during the growth of three *Origanum vulgare* groups.

populations in the flowering stage seems appropriate to the life cycle of the plant. Earlier collection causes slow regrowth with some degree of flowering, combined with death or weakening of the plants (7). On the other hand the optimum harvest date for a high yield of essential oil is in May, when the plants are at the early flowering stage. This result is in accordance with observations in other aromatic plants (8). The main constituents of the essential oil changed gradually during the maturity of the plant (6).

The taste and aroma of group A were less pungent than those of group B and C. Group A also had a sweet alcoholic character and resembled the taste and aroma of sweet marjoram with a phenolic pungency. Those variations could be used for cultivation, when the users become more familiar

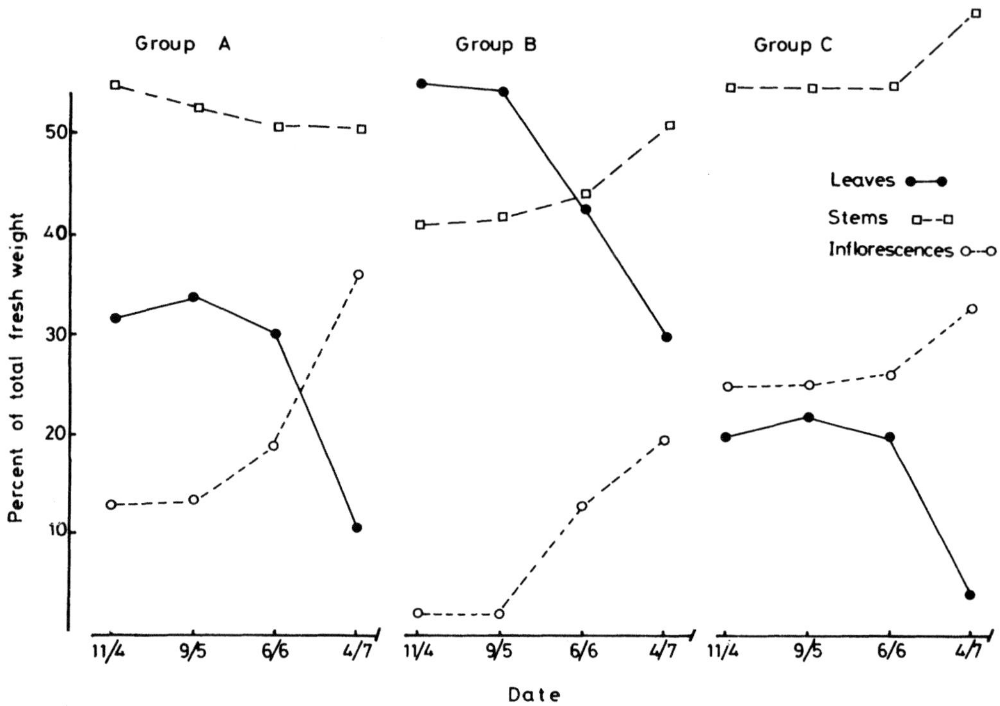

Fig. 2. Changes in various parts of the plant (percentage of total weight) during the growth of three *Origanum vulgare* groups.

Table I. Percentages of the essential oil and the main components during the growth of three groups of *Origanum vulgare* strains.

	Date of sampling	Percentages		
		Group A	Group B	Group C
Essential oil	11/4	4.0	3.8	2.2
	9/5	4.9	4.3	4.4
	6/6	4.5	4.0	3.5
	4/7	4.3	3.5	3.2
Phenols (thymol + carvacrol)	11/4	20.4	72.8	54.4
	9/5	20.6	76.6	55.5
	6/6	24.3	75.0	56.4
	4/7	26.2	76.7	56.5
Monoterpene hydrocarbons (p-cymene + γ-terpinene)	11/4	29.8	18.8	29.9
	9/5	34.9	19.2	28.5
	6/6	22.9	15.3	32.3
	4/7	22.7	11.3	32.6

with uniform high quality herbs instead of variable products collected from wild populations.

REFERENCES

(1) Putievsky E, Basker D. 1977, J. Hort. Sci. 52:181.
(2) Putievsky E. 1978, Hassadeh 58:1269 (in Hebrew).
(3) Greenhalgh P. 1979, The market for culinary herbs. Report of the Tropical Products Institute, No. G121, London.
(4) Parry, JW. 1962, Spices, their Morphology, Histology and Chemistry, New York, Chemical Publ. Co.
(5) Putievsky E, Ravid U. 1982, in: Aromatic Plants: Basic and Applied Aspects (Margaris N, Koedam A, Vokou D, Eds.), The Hague/Boston/ London, Martinus Nijhoff Publishers, p. 237.
(6) Basker D, Putievsky E. 1978, J. Hort. Sci. 53:183
(7) Putievsky E. 1985, Growing Aromatic Plants in Mountain Area without Irrigation, in preparation.
(8) Duhan SPS, Garry SN, Roy SK. 1975, Indian J. Pharm. 37:41.

COMPARATIVE STUDIES ON GROWTH AND VOLATILE OIL CONTENTS OF SOME INDUCED MUTANTS OF *ORIGANUM MAJORANA*

N.E. EL-KELTAWI, R.F. ABDOU, D.W. BISHAY

ABSTRACT

Marjoram (*Origanum majorana* L.) is used in medicinal as well as in perfumery industries. Seeds of marjoram contain fixed oil. Growers of aromatic plants usually use seed propagation for marjoram. This method leads to less essential oil production and to a lower quality and quantity of the herb. It may be due to the segregations of certain genetic characteristics of the plant. On the other hand, marjoram is easily vegetatively propagated with both cuttings and division.

Chemical mutagens are considered as a mean to produce mutations.Trimethanolamine (TMA), diethylamine (DEA) and diethyl sulphate (DES) were applied at concentrations of 0.005, 0.05, 0.25, 0.5, 1.0, 2.0 and 4.0%. Marjoram seeds were soaked in freshly prepared water solutions of these chemicals for 4 h at 25°C. Germination percentages were determined after one week. Seedlings were cultivated in pots of 20 cm diameter. Six mutants were selected and propagated with cuttings for two generations.

The data of two seasons showed the following results: generally, germination percentage and characteristic of plants differed significantly compared with the control; the selected mutants showed a wide variation concerning growth, flowering, essential oil contents, photosynthetic pigments in leaves, and rooting of cuttings; two mutants contained nearly a double essential oil percentage compared with the normal percentage of marjoram; one mutant contained no essential oil.

INTRODUCTION

Origanum majorana L., sweet marjoram, is indigenous to Mediterranean countries (1) and is cultivated in Central Germany, Hungary, Southern France and USA (2). The plant is cultivated in Egypt for its leaves, essential oil and seeds. The leaves are used as condiment and the essential oil in medicinal and perfumery industries (3). Marjoram seeds contain a fixed oil, which is used in confectionary (2).

Mostly, growers of aromatic plants have used seed propagation for marjoram. This method of propagation leads to less essential oil production and to a lower quality and quantity of the herb, as a result of the variations between the plants in the field. It may be due to the segregations of certain genetic characteristics of the plants. On the other hand, marjoram is easily vegetatively propagated with both cuttings and division.

A. Baerheim Svendsen and J.J.C. Scheffer (eds), Essential oils and aromatic plants, ISBN-13: 978-94-009-5137-2

Mutations are the source of genetic variability and evolution. Efforts were made to induce mutations in order to select new or entirely different characteristics, which might be of interest for breeding programs.

Chemical mutagens were used fourty years ago to induce mutations (4). Most of the biological effects induced in seeds by chemical mutagens can be observed as reduction of survival, seed germination percentage and seedling growth. Moreover, an increase in the number of chromosomal aberrations in root-tips, chlorophyll-deficiency mutant tissue chimeras and/ or mutation frequency were noticed (5-9).

The present experiments were aimed to induce mutants of high aromatic value in sweet marjoram by using trimethanolamine, diethylamine and diethyl sulphate as mutagens.

MATERIAL AND METHODS

The investigations were carried out for three successive seasons, 1981-82, 1982-83 and 1983-84, at the Experimental Farm of Floriculture of the Assiut University, that provided us with seeds of the local variety of sweet marjoram. Chemical mutagens were trimethanolamine (TMA), diethylamine (DEA) and diethyl sulphate (DES), obtained from Eastman Kodak.

The chemical mutagens were used at concentrations of 0.005, 0.05, 0.25, 0.5, 1.0, 2.0 and 4.0%, besides the control treatment (distilled water). In November 1981 800 seeds for each treatment were soaked in a freshly prepared solution at 25°C for 4 h. The half of the treated seeds was washed thoroughly with distilled water and immediately left for germination in Petri dishes for 1 week at 25°C. The germination percentage was calculated. The lay-out of this experiment was block randomized design with 4 replicates. The second half of the treated seeds were sown in seed pans containing Nile loam soil. They were placed in the shade and watered daily. After germination had taken place the seed pans were transferred to a sunny place to harden the seedlings; they were transferred to 20 cm pots in the beginning of February 1982. The marjoram plants were arranged in block randomized design with 14 treatments and 4 replicates. Each plot comprised 10 pots. At the end of May 1982, the plant height was determined. Samples of inflorescences of some mutants were collected and photographed. Six plants of the whole experiment were selected as quite different mutants according to the strength and sweetness of their odour, the colour of their leaves and the shape of their inflorescences.

The selected mutants and the control plants were replanted in 30 cm pots to obtain more growth for stem cuttings. In mid August 1982, stem cuttings of 15 cm were used for propagation. 50 cuttings of each mutant and the control were examined for rooting. Each 10 cuttings were planted in a 15 cm pot filled with peat moss. The pots were arranged in block randomized design with 5 replicates. Meanwhile vegetative propagation with stem cuttings was done until sufficient plants were obtained to carry out the experiment of comparing the mutants. After 1 month the rooting percentage was calculated

In the beginning of October 1982, the rooted cuttings were transplanted into 15 cm pots filled with clay soil. After 1 month the plants were replanted into 25 cm pots. The plants were arranged in block randomized design with 7 treatments and 4 replicates. Each plot comprised 6 pots. In mid March 1983, data on the plant height and total fresh weight for each mutant and the control plants were recorded. Photosynthetic pigments, chlorophyll a and b, carotenoids and total pigments, were determined in the fresh leaves according to Metzner (10). Essential oil percentages in the fresh leaves were determined according to the Egyptian Pharmacopoeia (11). The yield of essential oil per plant was calculated. The experiment was repeated at the same dates during the second season (1983/84). Average data of both seasons were statistically analyzed (12). In mid June 1984, the amounts of seeds per plant were recorded.

The essential oils of the different mutants and the control plants were subjected to qualitative and quantitative study using TLC. Silica gel G on glass plates was used as adsorbent and n-hexane-ethyl acetate (85:15) as solvent system. Vanillin sulphuric acid spray reagent (1%) was used for visualisation of the separated components.

RESULTS AND DISCUSSION

Effect of chemical mutagens

The data obtained on the germination percentage (see Table I), revealed that the treatment of marjoram seeds with TMA, DEA and DES solutions of 1.0, 0.25 and 0.005%, respectively, significantly reduced the germination percentage compared with the control. Concentrations over 0.005% of DES completely inhibited the seed germination. Generally, germination percentage was decreased with increasing mutagen concentration. The most effective mutagen on reduction of the germination percentage was in the following order DES > DEA > TMA at the same concentration. Similar results were reported by Abdou (13) on wheat and by Hildering (7) on tomato, by using ethylene imine.

Table I. Effect of chemical mutagens on seed germination and plant height of *Origanum majorana* L.

	Percentage of mutagens							
	0.005	0.05	0.25	0.50	1.0	2.0	4.0	Control
	Germination percentage							
TMA	92.3	88.8	85.6	82.3	76.8	51.3	20.3	96.8
DEA	90.3	89.0	73.8	58.0	23.0	7.3	0	
DES	82.8	0	0	0	0	0	0	
	Plant height in cm							
TMA	19.0	22.5	24.8	27.0	29.3	21.0	16.3	19.8
DEA	18.3	21.5	22.8	25.5	17.8	-	-	
DES	14.8	-	-	-	-	-	-	

LSD_{05} (germination percentage) = 21.08 LSD_{05} (plant height) = 1.6
LSD_{01} (germination percentage) = 16.01 LSD_{01} (plant height) = 1.2

Six months after sowing the seeds treated with 0.05-2.0% TMA and 0.05-0.5% DEA an improved height of the marjoram plants was observed (Table I), whereas a reduced plant height was observed with 0.005% DES and/or 1.0% DEA. Many authors (7, 13, 14) reported similar results on some plant species after treatment with certain chemical mutagens.

Many variations between the plants, particularly as to the growth and the shape of the inflorescences were seen as the effect of the mutagens on the phenotype of the plants produced from the treated marjoram seeds. Also variations in colour, thickness and size of the leaves were observed.

Study of selected mutants

Generally all mutants showed good rooting percentage. Apparently, the percentage of the rooted cuttings of the selected mutants, except No. 6, were significantly higher than that of the control during both years (Table II). This means that there is no problem for propagation of the selected mutants as new cultural varieties of marjoram. The intensity and the distribution of roots on the cuttings varied for the different mutants.

The total fresh weight of the mutants No. 2-6 was superior as compared with the control plants during both years (Table II). The maximum yield was obtained from mutant No. 6, whereas mutant No. 1 gave the lowest fresh weight as compared with the control and/or the other mutants.

The weight of the air-dried seeds per plant showed variable results (Table II). The maximum yield of seeds was obtained from mutant No. 6, the minimum from mutant No. 2. The percentage of oil distilled from the fresh herb of marjoram showed variable results for the different mutants (Table III). Mutants No. 2, 4 and 6 contained more essential oil compared to the control, whereas no significant differences were seen in the oil percentages of mutants No. 3 and 5. No essential oil was obtained from mutant No. 1.

Table II. Ability of the selected mutants on rooting, yields of herbage and seeds during the first and second year.

Mutant No.	Rooting percentage 1982/83	Rooting percentage 1983/84	Fresh weight in g per plant 1982/83	Fresh weight in g per plant 1983/84	Weight of seeds in g per plant 1983/84
1	87.8	88.8	87.3	88.5	3.43
2	85.5	87.0	237.0	234.0	2.73
3	78.5	80.0	257.0	259.5	5.98
4	70.0	74.3	231.0	216.8	2.85
5	86.5	85.8	254.5	242.5	5.70
6	62.5	63.8	280.8	285.3	7.35
Control	66.8	66.8	176.3	150.3	4.00
LSD_{05}	6.9	7.3	14.6	8.3	0.50
LSD_{01}	9.4	9.9	19.9	11.2	0.68

Table III. Essential oil percentage and oil yield per plant in fresh herb of the selected mutants of marjoram during the first and second year.

Mutant No.	Oil percentage (v/w) 1982/83	Oil percentage (v/w) 1983/84	Oil yield in ml per plant 1982/83	Oil yield in ml per plant 1983/84
1	0	0	-	-
2	0.73	0.75	1.73	1.75
3	0.49	0.48	1.28	1.25
4	0.76	0.78	1.76	1.67
5	0.43	0.43	1.08	1.04
6	0.62	0.61	1.66	1.74
Control	0.44	0.41	0.78	0.61
LSD_{05}	0.04	0.09	0.19	0.11
LSD_{01}	0.05	0.14	0.30	0.17

The data in Table III indicated that the oil yields of the different mutants were superior compared to the control during both years. The essential oil samples from the different mutants and from the control showed no difference in colour, but they varied in their odour. TLC of the different oils did not show any variation in the number of constituents compared with the control. So the difference in the oil odour may be due to quantitative variations of the components.

Many investigators have reported that marjoram species and varieties differ in their essential oil percentage and quality (2, 15).

The colour of the leaves is an important characteristic for the consumers or aromatic plants. The data in Table IV revealed that mutants No. 1,5 and 6 contained larger total amounts of pigments than the control. Some investigators (5, 6, 16) recorded that mutation is an important factor affecting chlorophyll accumulation.

Table IV. Amounts of pigments (mg/g fresh weight) in leaves of the selected mutants of marjoram

Mutant No.	Chlorophyll a	Chlorophyll b	Carotenoids	Total pigments
1	9.71	2.16	4.43	16.30
2	5.29	1.59	3.68	10.56
3	6.13	0.78	3.58	10.49
4	5.50	0.44	3.69	9.63
5	10.24	1.81	4.56	16.61
6	11.59	2.23	5.72	19.54
Control	6.13	0.80	3.57	10.50
LSD_{05}	1.08	0.25	0.33	1.02
LSD_{01}	1.47	0.34	0.45	1.39

It could be concluded that mutant No. 6 is of high value for the production of essential oil, leaves and seeds, and it is easily propagated by cutting. So it can be used as a new cultural variety.

REFERENCES

(1) Parry ED. 1925, Parry's Cyclopaedia of Perfumery, London, JA Churchill Ltd.
(2) Guenther E. 1961, The Essential Oils, New York/London, D. van Nostrand Comp., Vol. 3.
(3) Chopra RN, Nayar SL, Chopra IC. 1956, Glossary of Indian Medicinal Plants, New Delhi, Council of Scientific & Industrial Research.
(4) Elliot FC. 1958, Plant Breeding and Cytogenetics, New York, McGraw-Hill Book Comp. Inc., p. 125.
(5) Heiner RE, Konzak CF, Nilan RA, Legault RR, 1960, Proc. Nat. Acad. Sci. 46:1215.
(6) Konzak CF, Nilan RA, Harle JR, Heiner RE. 1961, Control of factors affecting the response of plants to mutagens. Brookhaven Symposia in Biology, 14:128.
(7) Hildering GJ. 1962, Euphytica 12:113.
(8) Favret EA. 1959, Mutation research in crop plants. Eucarpia, 2nd Congress, Cologne, FRG, p. 76.
(9) Nilan RA, Konzak CF. 1961, Increasing the efficiency of mutation induction. Mutation and Plant Breeding. NAS-NRC 891:437.
(10) Metzner H, Ran H, Senger H. 1965, Planta 65:185.
(11) Egyptian Pharmacopoeia. 1972, Cairo, Cairo University Press.
(12) Snedecor GW. 1956, Statistical Methods, Ames, Iowa, USA, The Iowa State College Press, 5th Ed.
(13) Abdou RF. 1974, Thesis, Faculty of Agriculture, Assiut University, Egypt.
(14) Murphy CF, Patterson FL. 1958, The effect of the chemical mutagen, diepoxybutane on Clint-land Oats. Proc. Iowa Acad. Sci. 165:184.
(15) Lossner G. 1967, Pharmazie 22:324.
(16) Rebeiz CA, Calstelfranco PA. 1973, Ann. Rev. Plant Physiol. 24:129.

FORMATION OF ESSENTIAL OIL IN CLARY SAGE UNDER DIFFERENT CONDITIONS

G. VERZÁR-PETRI, M. THEN, S. MÉSZÁROS

ABSTRACT

Some factors of importance for the quality of the essential oil of *Salvia sclarea* are discussed. The effect of environmental conditions and distillation parameters on the oil quality is pointed out. Under controlled conditions, a standard quality of the essential oil can be achieved.

INTRODUCTION

Plants, chiefly in the tropics and the sub-tropics, produce essential oils in large quantity. Also in Hungary the climatic and soil conditions are suitable for cultivation of some medicinal plants and spices containing essential oils.

The industrial distillation of essential oils takes place in many big farms directly from plants cultivated there. The high price of the various oils requires standard quality that can be achieved when good technological processes are used. The plant material, its developing stage, the processing before and after harvesting, as well as large-scale distillation (1-3), are important factors for giving a good quality of the oil. The aim of the work described here, was to find optimum agricultural and technological parameters for producing high quality essential oils. The influence of external conditions (climate, soil) were studied in order to obtain the best harvest. The investigations were performed in laboratory and in field scales, varying the comminution of the plant material and the amounts of it in the distillation apparatus.

MATERIAL AND METHODS

Clary sage, *Salvia sclarea* L., was used in all experiments.

The analysis of the essential oils was carried out with volumetric as well as physical methods. Quality control was performed with GC using a gas chromatograph Jeol 1100 equipped with a glass column (2 m x 3.4 mm)

A. Baerheim Svendsen and J.J.C. Scheffer (eds), Essential oils and aromatic plants, ISBN-13: 978-94-009-5137-2

packed with 3% OV-17 on Chromosorb W; other conditions: oven temperature, 100-250°C (4°C/min); carrier gas, Nitrogen. For some experiments capillary GC was also applied (4).

RESULTS AND DISCUSSION

Salvia sclarea is a biennial plant in Central Europe. Sporadically it flowers in the first year, but usually only in the second year. Then the first flowering period is in June-July and the second one in September. The linalyl acetate content that characterizes the fragrance of clary sage oil is highest during the first flowering period in the second year. The ester value, calculated as linalyl acetate and used for the standardization of the oil may be as high as 70% (Table I). Ester values of French and English oils are 54-70% and 34-53% respectively. The free linalool, 1.8-cineole and limonene contents are also characteristic factors with their maxima in the second year as well.

The colour of the oil is an important factor in the trade. The distillation can result in a light-coloured oil when the plant material is harvested in the first period of the second year. A higher percentage of leaves in the raw material yields a darker oil.

The essential oil composition in leaves differs from that in flowers. Main constituents of leaf oil are β-thujone, 1.8-cineole and terpenic acid; bornyl acetate is a characteristic ester.

Studying the essential oil composition in *Salvia sclarea* during the second flowering period we found that the contents of β-thujone and borneol are higher in the full-flowering stage than in the bud stage. Increase of the ester content takes place in the full-flowering stage. The composition of the oil changes during the whole period of harvest and distillation. The linalyl acetate content decreases often, α- and β-pinene disappear; 1.8-cineole and myrcene decrease, influencing the scent of the oil. In the first period of harvest a substantial higher linalyl acetate content was observed (5).

However, the linalyl acetate content varies markedly and is influenced by external conditions. In addition, changes in the alcohol solubility and optical rotation also deserve attention. The composition of essential oil samples produced in various districts of Hungary, i.e. Biatorbágy, Daránypuszta, Herceghalom and Páty, varied to some extent.

Table I. Characteristics of different clary sage essential oil samples.

District of cultivation	Soil type	Colour of oil	Density (temperature)	n_D^{20}	Optical rotation	Solubility in alcohol 98%	Percentage of total esters	Percentage of linalyl acetate
Darānypuszta$_1$	hard clay	brown	0.901 (20°C)	1.470	- 8°	1.2	60.3	42.3
Darānypuszta$_2$		brown	0.899 (30°C)	1.4629	-11.1^o	0.8	65.1	45.58
Herceghalom$_1$	limy	green	0.8971 (20°C)	1.4662	-11.6^o	0.57	90.57	63.33
Herceghalom$_2$		yellow	0.896 (20°C)	1.473	-11.2^o	1.01	67.7	47.41
Pāty$_1$	sand	yellow	0.9031 (15°C)	1.46	-10.6^o	0.47	93.4	65.35
Pāty$_2$		yellow	0.8998 (20°C)	1.4711	-55.7^o	0.11	72.01	53.9
Biatorbāgy	humus	dark-yellow	0.989 (20°C)	1.4870	0°	0.86	59.5	41.7
Commercial oil sample		yellow	0.899 (19°C)	1.4690	-25.6^o	0.38	69.4	48.87

Finely cutting of the plant material resulted in dark oils with a low ester content. The oil yield in distillation apparatuses with much plant material is lower than that in apparatuses containing less material (Table II). In laboratory-scale experiments the essential oil composition of plant material chopped up into pieces of two different sizes, 2-3 mm and 8-10 mm respectively, was investigated. Three fractions were collected from each sample during a distillation period of 2 h.

The main factors for an optimum production of the essential oil from clary sage are the amount of plant material in the distillation apparatus and the distillation period.

Table II. Influence of the amount of plant material in the distillation apparatus on the oil yield in percentages.

Amount of plant material in kg	Yield of oil	Main constituent of oil
0.650	3	Linalyl acetate
1.500	2	Linalool

REFERENCES

(1) Erickson RE. 1976, Lloydia 39:8.
(2) Gildemeister E, Hoffmann Fr. 1961, Die ätherischen Öle, Band VI, Berlin, Akademie-Verlag.
(3) Hornok L. 1968, Cultivation and Processing of Medicinal Plants, Budapest, Mezögazdasági Kiadó, p. 80.
(4) Verzár-Petri G. 1981, Szabványositás (Standardization) 6:184.
(5) Then M. 1977, Ph.D. Thesis, University of Budapest, Hungary.

INFLUENCE OF ECOLOGICAL FACTORS ON THE CONTENT AND COMPOSITION OF THE THE ESSENTIAL OIL IN *SALVIA OFFICINALIS*

I. PITAREVIĆ, D. KUŠTRAK, J. KUFTINEC, N. BLAŽEVIĆ

ABSTRACT

The quality of sage leaves (content and composition of the essential oil) collected each month from November 1982 till May 1983 on two localities, Kuna (700 m altitude) and Doli (200 m) was examined. In the same period meteorological data were recorded. The amount of oil was determined by hydrodistillation, and the composition was analyzed by GC.

INTRODUCTION

The summer 1982 was characterized by very dry and hot weather, and it resulted in retarded vegetation and slow development of sage leaves. The condition of the plant material influenced the harvest of sage leaves in the district of Dubrovnik. In the autumn, i.e. in September and especially in October a rainy period with a lot of precipitation followed, leading to an intensive growth of the sage plants. We collected samples of sage leaves and evaluated their quality with respect to the amount and composition of the essential oil. This report is a part of our extensive investigations on the yield and composition of Dalmatian sage oil (1-2).

MATERIAL AND METHODS

Samples of wild growing Dalmatian sage (*Salvia officinalis* L.) were collected at two localities in the Dubrovnik district: Kuna (Konavle) is on the sea-side slope of mount Snjeznica at an altitude of 700 m; Doli is located west of Dubrovnik near the city of Ston at an altitude of 200 m.

The amount of essential oil was determined using the apparatus of the Yugoslavian pharmacopoeia (3); 50 g of sage leaves 'in toto' was taken and the distillation lasted for 3.5 h. Refractive index and content of ketones calculated as thujone were determined according to the procedure given in the same pharmacopoeia for Salviae aetheroleum (hydroxylamine method).

Before the introduction of GC, the quality of sage oil was evaluated

A. Baerheim Svendsen and J.J.C. Scheffer (eds), Essential oils and aromatic plants, ISBN-13: 978-94-009-5137-2

on the content of ketones, calculated as thujone. According to the pharmacopoeia mentioned Salviae aetheroleum should contain not less than 40% of ketones. We analyzed the oil samples also by GC. A gas chromatograph Pye Unicam 204 equipped with FID and a home made Pluronic 68 WCOT glass capillary column (30 m) was used; oven temperature, 70°C (5 min), then 70-200°C (3°C/min) and then isothermal 4 min; injector and detector 250°C; carrier gas, Hydrogen.

The meteorological data were collected at the meteorological station in Ston for the period June 1982-May 1983.

RESULTS AND DISCUSSION

Dalmatian sage is a xerothermophylic species and it grows spontaneously on rocky, extremely arid soil. Investigating the samples from two different localities, we noticed significant differences in the amount of essential oil in the leaves and in the percentages of the various oil components. The amount of oil in the samples from Kuna was higher than that in the samples from Doli in November-January (Table I).

Table I. The amount in percentage of essential oil, and its total content of thujone in *Salvia officinalis* collected from November 1982 till May 1983. * No collection because of the weather.

Sample		Date of harvest		Essential oil % (v/w)	Refractive index (20°C)	Total thujone %
				Kuna (Konavle)		
1	1982	November	10	2.50	1.4668	42.04
2		December	21	1.82	1.4650	41.07
3	1983	January	25	1.50	1.4623	48.78
4		February	*			
5		March	30	1.24	1.4632	47.30
6		April	20	1.31	1.4655	36.50
7		May	23	2.20	1.4652	27.81
				Doli (Dubrovnik)		
1	1982	November	20	1.90	1.4645	51.42
2		December	22	1.50	1.4620	50.90
3	1983	January	24	1.32	1.4626	52.51
4		February	24	1.24	1.4624	48.17
5		March	29	1.36	1.4615	51.82
6		April	25	1.40	1.4664	46.36
7		May	24	2.42	1.4630	41.32

The maximum absolute temperature in the Kuna region in the summer is 40°C. Therefore, the rocks there accumulate more heat so that the ground keeps a high temperature longer than at Doli. Because of heavy precipitation and low temperatures at Kuna in February, we were not able to collect samples there in that month. In March-April the amount of essential oil was higher at Doli, where the soil was warmed up earlier than at Kuna. In Table II the temperature and the amount of rainfall are shown. In November and December the temperatures were relatively high and much precipitation was noted.

Table II. Meteorological data during the sampling period.

Month		Temperature in °C Maximum	Minimum	Rainfall in mm	Number of rainy days
1982	June	32.3	12.0	6.1	2
	July	34.0	14.7	56.2	6
	August	33.5	13.5	60.1	6
	September	33.0	11.0	210.9	5
	October	28.5	6.0	385.6	15
	November	22.0	3.0	178.2	7
	December	18.0	-2.0	312.0	16
1983	January	12.7	1.7	13.5	3
	February	10.5	1.3	116.1	12
	March	15.3	4.9	87.3	9
	April	18.5	7.3	19.3	6
	May	23.3	11.3	25.9	6

A long dry season should give a high oil production (4). It can explain the relatively high content of essential oil in November at both localities (2.5% and 1.9%) after the dry and hot months June-August.

The seasonal variation of the main compoments is illustrated in the Figs. 1-3. In the samples investigated, the content of thujone was always higher in the samples from Doli than in those from Kuna (Fig. 3).

Vernazza (5) emphasized that the refractive index of sage oil highly depends on its thujone content: the higher the percentage of thujone, the lower the refractive index. We found this to be correct also for our samples (Table I).

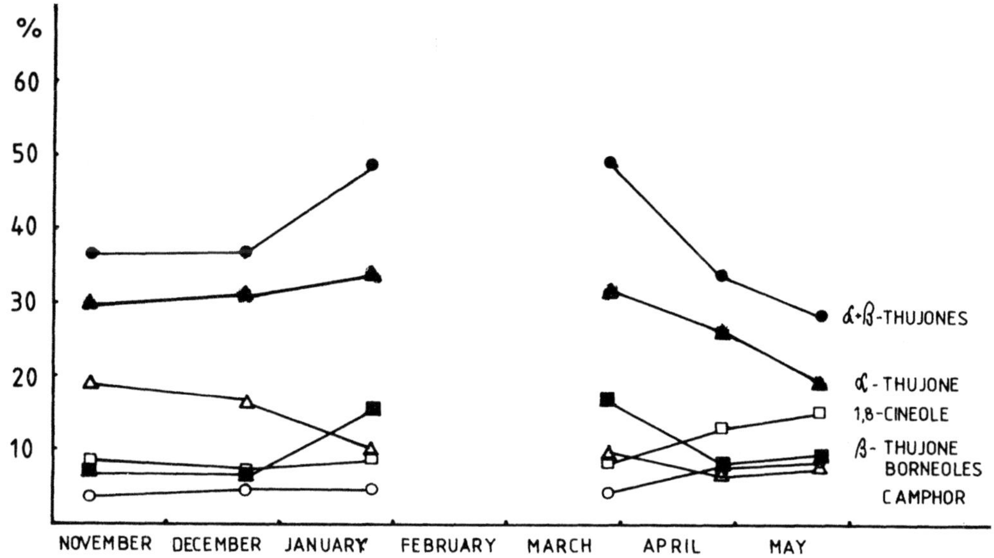

Fig. 1. Seasonal variation of the main components of Dalmatian sage oil from Kuna.

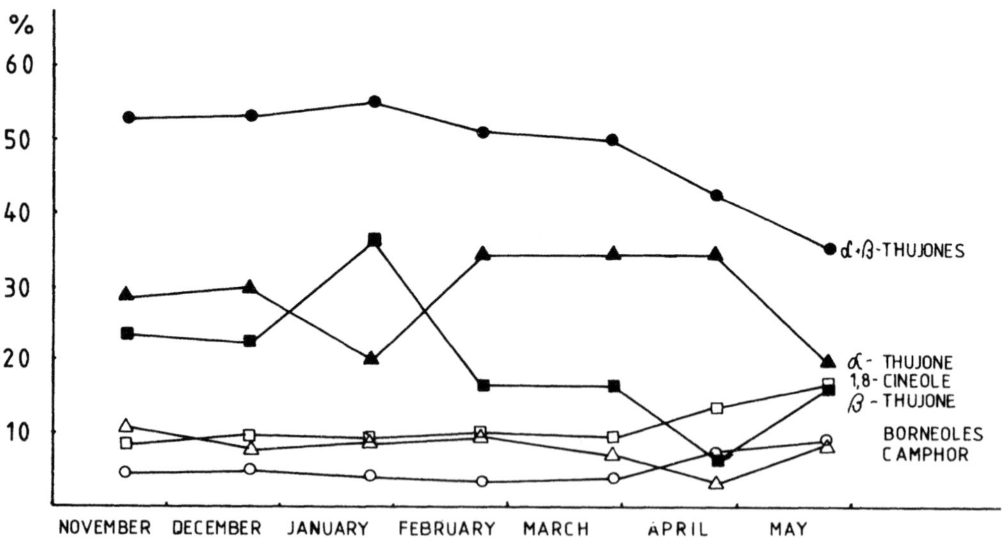

Fig. 2. Seasonal variation of the main components of Dalmatian sage oil from Doli.

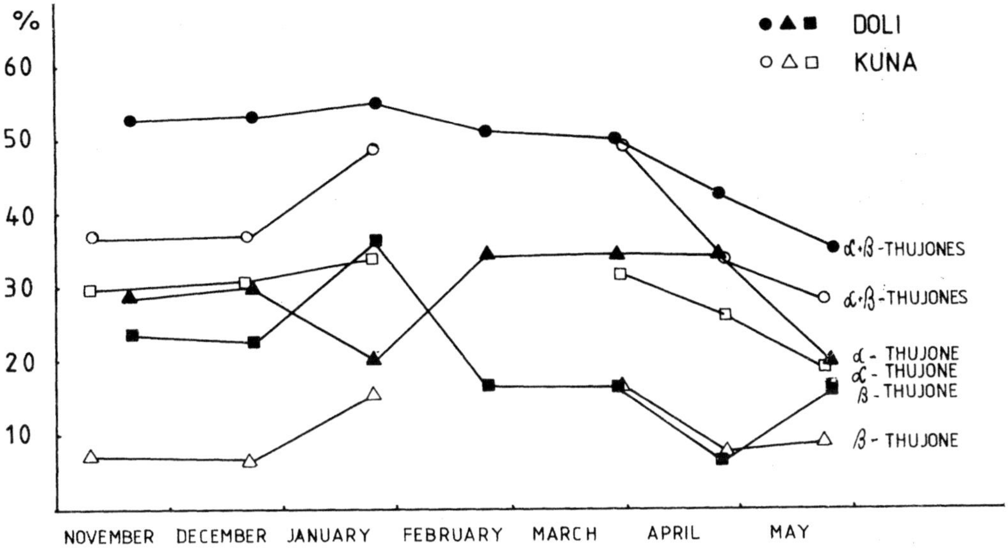

Fig. 3. Seasonal variation of α- and β-thujone in Dalmatian sage oil from Kuna and Doli.

REFERENCES

(1) Pitarevič I, Kuftinec J, Blaževič N, Kuštrak D. 1984, J. Nat. Prod. 47: 409.
(2) Kuštrak D, Kuftinec J, Blaževič N. 1984, J. Nat. Prod. 47:520.
(3) Pharmacopoea Jugoslavica, Editio tertia, Vol.I. Beograd, 1972.
(4) Burmeister J, Von Guttenberg H. 1960, Planta Med. 8:1.
(5) Vernazza N. 1957, Acta Pharm. Jugoslav. 7:163.

COMPOSITION OF THE ESSENTIAL OIL OF *ORIGANUM MAJORANA* GROWN IN DIFFERENT LOCALITIES IN TURKEY

E. SARER, J.J.C. SCHEFFER, A.M. JANSSEN, A. BAERHEIM SVENDSEN

ABSTRACT

Origanum majorana, wild growing in Turkey, was collected in three localities. The corresponding essential oils were investigated by GC and GC-MS. The contents of carvacrol and thymol varied from 48-74% and 0.5-4%, respectively. The oils were also tested against bacteria, a yeast and some fungi. A remarkable effect against the fungi used in the screening was observed.

INTRODUCTION

Origanum species have been used in medicine and as spices ever since antiquity – mainly because of their content of essential oils. *Origanum majorana* L. (synonyms: *Majorana vulgaris* Miller, *Majorana hortensis* Moench, *Origanum dubium* Boissier) is a native of Cyprus and the adjacent part of Southern Turkey (1). In a previous study (2) we investigated the essential oil isolated from flowering parts of *O. majorana* plants grown near Alanya (Southern Turkey) along the coast. The oil contained large amounts of carvacrol. The high percentage of phenolic compounds was interesting in view of literature data on the species concerned. For this reason material from two other localities was collected and subsequently analyzed. In the present paper the composition of the two oil samples isolated from this material is compared with that of the oil sample studied before (2). The antimicrobial activity shown by the oil samples and by some of their components is also described.

MATERIAL AND METHODS

Flowering parts of *Origanum majorana* L. were collected near Alanya, as the material (= sample 1) studied previously, but at an altitude of 1000 m (= sample 2), and near Içel some hundred kilometers east of Alanya (= sample 3). The plant material collected in both localities was dried at room temperature and subjected to hydrodistillation for 3 h using a

A. Baerheim Svendsen and J.J.C. Scheffer (eds), Essential oils and aromatic plants, ISBN-13: 978-94-009-5137-2

Clevenger-type apparatus. The essential oils were compared by means of GC analysis. Comparison with authentic samples and GC-MS were used for identification of the oil constituents. The capillary GC and GC-MS were performed as described elsewhere (3).

In addition the essential oil samples were tested for activity against some bacteria, a yeast and some fungi. The antimicrobial screening was carried out by means of the agar overlay technique (3).

RESULTS AND DISCUSSION

The composition of the essential oil of *O. majorana* (sample 1) that was studied previously (2), is shown in the Tables I and II. This sample was analyzed again under the present analytical conditions, and the results were compared with those obtained in 1982. The oil composition was also compared with that of the samples 2 and 3, and striking differences are given in Table III.

Table I. Composition of the mixture of monoterpene hydrocarbons from the essential oil of *Origanum majorana* (sample 1). The mixture accounts for 13% of the total oil; tr = trace (< 0.05%).

Compound	%	Compound	%
Tricyclene	tr	α-Terpinene	7.4
α-Pinene α-Thujene	9.2	Limonene	1.4
		β-Phellandrene	2.9
Camphene	1.7	cis-β-Ocimene	1.0
β-Pinene	1.0	γ-Terpinene	24.8
Δ4-Carene	0.1	trans-β-Ocimene	0.5
Δ3-Carene	0.4	p-Cymene	36.0
Myrcene α-Phellandrene	10.6	Terpinolene	0.7

Table II. Composition of the mixture of oxygen-containing compounds (exclusive phenols) from the essential oil of *Origanum majorana* (sample 1). The mixture accounts for 18% of the total oil; tr = trace (< 0.05%).

Compound	%	Compound	%
1.8-Cineole	1.4	Bornyl acetate	tr
3-Octanol	0.3	Terpinen-4-ol	3.6
Fenchone	1.5	cis-p-Menth-2-en-1-ol	0.2
trans-Sabinene hydrate	2.7	Isoborneol	tr
β-Thujone	0.3	α-Terpineol	3.3
Menthone	tr	Borneol	2.8
Linalool	76.6	Carvone	0.6
cis-Sabinene hydrate	1.6	Geraniol	0.2
trans-p-Menth-2-en-1-ol	0.5		

Table III. Differences in the composition of the essential oil samples 1, 2 and 3 of *Origanum majorana*. For sample 1 the figures (percentages of the total oil) obtained under the present analytical conditions and those obtained two years before are given.

Compound	Sample 1 1982	Sample 1 1984	Sample 2	Sample 3
γ-Terpinene	3.1	3.4	5.9	5.7
p-Cymene	4.5	7.0	3.5	4.8
Linalool	14.1	17.5	28.9	4.4
Thymol	3.9	2.2	3.6	0.5
Carvacrol	65.1	58.3	48.4	73.5

The results of the antimicrobial screening are shown in Table IV. A remarkable effect of the oil samples and some of their main components was observed, in particular against some fungi that are known to cause skin infections in humans. One should bear in mind that the Table shows inhibition diameters which have no direct correlation with the antimicrobial activity expressed as minimum inhibitory concentration (MIC) values.

Table IV. The average (N ≥ 2) inhibition zone (in mm, including discs of 6 mm) for oil samples of *Origanum majorana* and for some of the main components. *Bs* = *Bacillus subtilis*; *Sa* = *Staphylococcus aureus*; *Ec* = *Escherichia coli*; *Pa* = *Pseudomonas aeruginosa*; *Ca* = *Candida albicans*; *Tc* = *Trichophyton concentricum*; *Tm* = *T. mentagrophytes*; *Tr* = *T. rubrum*; Ø of Petri dishes = 88 mm.

	Bs	*Sa*	*Ec*	*Pa*	*Ca*	*Tc*	*Tm*	*Tr*
Oil sample 2	34	32	25	8	40	88	88	88
Oil sample 3	31	40	25	8	32	88	88	88
γ-Terpinene	10	10	9	0	11			
p-Cymene	7	7	0	0	8	7	8	8
Linalool	8	8	10	0	11	12		12
Thymol	35	32	21	7	29	88	88	88
Carvacrol	32	33	25	8	40	88	88	88

Although 1 kg of dried material was available from both localities, further studies are necessary to investigate whether the amounts of linalool and phenols are related to the occurrence of different chemotypes.

ACKNOWLEDGEMENTS

We wish to record our thanks to Dr. J.H. Ietswaart, Free University Amsterdam, for identification of the plant material, and Mrs. A. Looman for technical assistance.

REFERENCES

(1) Ietswaart JH. 1980, A Taxonomic Revision of the Genus Origanum (Labiatae), The Hague/Boston/London, Leiden University Press.
(2) Şarer E, Scheffer JJC, Baerheim Svendsen A. 1982, Planta Med. 46:236.
(3) Janssen AM, Scheffer JJC, Baerheim Svendsen A, Aynehchi Y. 1984, Pharm. Weekbl. (Sci.) 6:157.

COMPOSITION AND ANTIMICROBIAL ACTIVITY OF THE ESSENTIAL OIL OF *DUCROSIA ANETHIFOLIA*

A.M. JANSSEN, J.J.C. SCHEFFER, A. BAERHEIM SVENDSEN, Y. AYNEHCHI

ABSTRACT

The essential oil from herb of *Ducrosia anethifolia* (DC.) Boiss., growing wild in Iran, was investigated by combined LSC and GC, and GC-MS. The oil consisted mainly of aliphatic compounds. α-Pinene, myrcene and limonene were the main components of the hydrocarbons that constituted 13% of the oil. n-Decanal, n-dodecanal, n-decanol, trans-2-dodecenal and cis-chrysanthenyl acetate were the major oxygen-containing constituents, which accounted for 87% of the oil. The aliphatic components can be held responsible for the remarkable activity of the herb oil against gram-positive bacteria, *Candida albicans*, and some dermatophytes. The antimicrobial screening was carried out by means of the agar overlay technique.

INTRODUCTION

Ducrosia anethifolia, family Apiaceae, is native to Afghanistan, Iran, Iraq and Pakistan. Recently the generic limits of the genus Ducrosia were reestablished (1). In Iran the plant is used to improve the smell of foods and drinks. In traditional medicine it is used to treat catarrh, headache and backache; in Karoon the seeds are given to children as an infusion in case of colic (2).

In a general screening for antimicrobial activity the essential oils isolated from herb and fruits of *D. anethifolia* were tested. The herb oil showed a remarkable activity against some fungi that are known to cause skin infections in humans, whereas the fruit oil exerted only a small effect. The oils were also tested for activity against bacteria and a yeast. In the present paper we describe the composition of the herb oil of the plant, and the antimicrobial activity shown by the oil and by some of its main components.

MATERIAL AND METHODS

Fresh leaves and stems of *Ducrosia anethifolia* (DC.) Boiss., growing wild in Karaj (40 km west of Tehran), were collected in May-June 1979; fruits were collected in July 1979 in the same locality. The essential

A. Baerheim Svendsen and J.J.C. Scheffer (eds), Essential oils and aromatic plants, ISBN-13: 978-94-009-5137-2

oil from the leaves and stems was isolated by hydrodistillation, for 2 h, using a Clevenger-type apparatus. The average yield was 2% (v/w) of oil. The fruits yielded 1.5% of oil after hydrodistillation for 6 h.

LSC over silica deactivated by addition of 5% water (3) was used to separate the hydrocarbons from the oxygen-containing compounds, and secondly to fractionate the oxygen-containing components.

Capillary GC was performed using a gas chromatograph Packard 428. Conditions: columns, fused silica (50 m x 0.23 mm i.d. and 25 m x 0.22 i.d.) coated with CP-Wax 57 cb (0.19 μm) and CP-Sil 5 cb (0.11 μm) respectively; oven temperature, 70°C (hydrocarbons) and 115°C (oxygen-containing components); injector, 200°C; detector (FID), 200°C; carrier gas, Nitrogen, 1.2 ml/min. The samples were injected using the split sampling technique (ratio ca. 1:100). Further details of the analytical procedures are given elsewhere (4).

In the screening for antimicrobial activity the following micro-organisms were used: *Bacillus subtilis* and *Staphylococcus aureus* (gram-positive bacteria); *Pseudomonas aeruginosa* and *Escherichia coli* (gram-negative bacteria); *Candida albicans* (yeast); *Trichophyton mentagrophytes* var. *interdigitale*, *T. rubrum* and *Epidermophyton floccosum* (fungi). The bacteria and the yeast were brought in the growing phase by incubation in a liquid medium. After the incubation period the medium was diluted (1:20) with physiologic salt solution (in case of *E. coli*, 1:40), and the test plates were inoculated with this suspension. For the fungi a different technique was used. A Sabouraud agar plate was inoculated with the fungus. This plate was then incubated until a thick mycelium was formed. The mycelium was brought into 15 ml of physiologic salt solution containing 0.001% of Tween 80 by means of a transfer loop. This suspension was vigorously shaken and applied to the test plates. Further details are given in Table I.

Table I. Methods of culture for the test organisms.

Micro-organism	Culture conditions	Test conditions
Bacteria	18 h at 37°C in TSB (5)	24 h at 37°C on TSB agar
Yeast	18 h at 37°C in TSB (5)	24 h at 37°C on MB agar
Fungi	7 days at 28°C on Sabouraud agar (6)	7 days at 28°C on Sabouraud agar

The agar overlay technique (7) was used for the antimicrobial assay. In Petri dishes (Ø = 88 mm) filled with the adequate medium and seeded with the test organism a filter paper disc (Ø = 6 mm) was placed. The disc was impregnated with 5 µl of the oils or the compounds to be tested. After the incubation period indicated in Table I the total diameter of the inhibition zone, if present, was measured. In order to check the sensitivity of the test organisms, 5 µl of the following solutions of antibiotics were applied to the discs: streptomycin 10 mg/ml in water (bacteria); nystatin 50 mg/ml in water (yeast); griseofulvin 50 mg/ml in acetone (fungi).

RESULTS AND DISCUSSION

The percentage composition of the fraction of oxygen-containing components from the herb oil of *Ducrosia anethifolia* is summarized in Table II. By GC of the total oil and the LSC fractions the total percentages of hydrocarbons (13%) and oxygen-containing components (87%) could be computed. Sesquiterpene hydrocarbons were hardly present in the herb oil.

Table II. Composition of the fraction of oxygen-containing compounds from the herb oil of *Ducrosia anethifolia*. tr = trace (< 0.05%).

Compound	%	Identification		
1.8-Cineole	tr	GC		
n-Octanal	0.2	GC	LSC	
n-Nonanal	0.7	GC	LSC	
Fenchone	0.2	GC	LSC	
Citronellal	0.2	GC	LSC	
n-Decanal	52.2	GC	LSC	MS
Linalool	0.3	GC	LSC	
n-Octanol	0.3	GC	LSC	MS
cis-Chrysanthenyl acetate	5.4	GC	LSC	MS
trans-2-Decenal	0.2	GC	LSC	MS
n-Nonanol	0.4	GC	LSC	MS
Unknown	0.4		LSC	MS (4)
Unknown	0.5		LSC	MS (4)
Unknown	0.4		LSC	MS (4)
n-Dodecanal	20.5	GC	LSC	MS
α-Terpineol	0.4	GC	LSC	MS
Geranial	0.2	GC	LSC	
n-Decanol	6.3	GC	LSC	MS
Geranyl acetate	tr	GC	LSC	
Citronellol	0.5	GC	LSC	MS
n-Undecanol	tr	GC	LSC	MS
Geraniol	tr	GC	LSC	
trans-2-Dodecenal	5.9	GC	LSC	MS
n-Dodecanol	0.8	GC	LSC	MS
n-Decanoic acid	1.8	GC	LSC	MS
n-Dodecanoic acid	0.3	GC	LSC	MS

The results of the antimicrobial assay are shown in Table III. The oils and the compounds tested are mainly active against the gram-positive bacteria, the yeast and the fungi; all of these compounds contributed to the effect of the oil. It is not uncommon that aliphatic aldehydes and alcohols are active against fungi (8, 9). Our investigation shows no marked difference in inhibition between n-decanol and n-dodecanol for bacteria, whereas in the case of the yeast and the fungi the inhibition zones of these compounds are different. One should bear in mind that Table III shows inhibition diameters which have no direct correlation with the antimicrobial activity expressed as minimum inhibitory concentration (MIC) values.

Table III. The average ($N \geq 3$) inhibition zone (in mm, including discs of 6 mm) for oils of *Ducrosia anethifolia*, for some of the main components and for solutions of antibiotics. *Bs* = *Bacillus subtilis*; *Sa* = *Staphylococcus aureus*; *Pa* = *Pseudomonas aeruginosa*; *Ec* = *Escherichia coli*; *Ca* = *Candida albicans*; *Tm* = *Trichophyton mentagrophytes*; *Tr* = *T. rubrum*; *Ef* = *Epidermophyton floccosum*.

	Bs	*Sa*	*Pa*	*Ec*	*Ca*	*Tm*	*Tr*	*Ef*
Herb oil	70	29	0	0	63	84	88	88
Fruit oil	7	7	0	0	7	26	24	23
n-Decanal	25	16	0	0	88	57	87	86
n-Dodecanal	17	18	0	0	9	56	60	78
n-Decanol	15	17	0	0	88	87	65	65
n-Dodecanol	8	14	0	0	10	13	12	9
trans-2-Dodecenal	60	25	0	7	13	71	86	78
Streptomycin	21	17	16	16				
Nystatin					13			
Griseofulvin						20	26	32

REFERENCES

(1) Alava R. 1975, Notes R. Bot. Gard. Edinburgh 34:183.
(2) Parsa A. 1959/1960, Medicinal Plants and Drugs of Plant Origin in Iran (III), Qualitas Plantarum et Materiae Vegetabilis, p. 139.
(3) Scheffer JJC, Koedam A, Baerheim Svendsen A. 1976, Chromatographia 9:425.
(4) Janssen AM, Scheffer JJC, Baerheim Svendsen A, Aynehchi Y. 1984, Pharm. Weekbl.(Sci.) 6:157.
(5) Verpoorte R, Tjin A Tsoi A, Van Doorne H, Baerheim Svendsen A. 1982, J. Ethnopharmacol. 5:221.
(6) Funder S. 1953, Practical Mycology, Oslo, Brøggers Boktr. Forlag, p.33.
(7) Barry AL. 1974, in: Current Techniques for Antibiotic Susceptibility Testing (Balows A, Ed.), Springfield, Ill. Charles C. Thomas.
(8) Kurita N, Miyaji M, Kurane R, Takahara Y, Ichimura K. 1979, Agric. Biol. Chem. 43:2365.
(9) Kurita N, Miyaji M, Kurane R, Takahara Y. 1981, Agric. Biol. Chem. 45:945.

EFFECTS AND SIDE-EFFECTS OF ESSENTIAL OILS

H. SCHILCHER

ABSTRACT

Essential oils are not only very complex composed but also chemically very heterogeneous. No wonder, that many different effects are known. Pharmacological and clinical tests and experience of the traditional medicine show that essential oils have the following effects on external application: hyperaemic, antiinflammatory, antiseptic, granulation stimulating, deodorizing, insecticide and repellent actions. On oral application the following effects are observed: expectorating, appetite stimulating, choleretic, cholekinetic, carminative, spasmolytic, antiinflammatory, antiseptic, diuretic, sedative and circulation stimulating. Many other effects are reported in the traditional medicine.

The main side-effects of essential oils are allergic reactions, some oils have phototoxic effects, only a few essential oils show necrotic, narcotic, nephrotoxic, hepatotoxic and cancerogenic actions. In many cases the side-effects are purely toxic effects caused by misuse of essential oils.

INTRODUCTION

Since ancient times essential oils have been used in therapy and as spices. However, since essential oils usually are very complex mixtures of mono- and sesquiterpenes as well as phenylpropane derivatives (hydrocarbons and several types of oxygen-containing compounds) the various essential oils have often quite different effects. Some of them are summarized in Table I.

Table I. Effects and side-effects of essential oils.

External application	Internal application
- hyperaemic	- expectorating
- antiinflammatory	- appetite stimulating
- antiseptic/disinfectant	- choleric, cholekinetic, carminative
- granulation stimulating	- antiseptic/disinfectant
- deodorizing	- sedative
- insecticide/insect repellent	- circulation stimulating

A. Baerheim Svendsen and J.J.C. Scheffer (eds), Essential oils and aromatic plants, ISBN-13: 978-94-009-5137-2

EXTERNAL APPLICATION

Because of the lipoid solubility of essential oils they are easily resorbed via skin and mucosa. Already in 1940 Strähli (1) examined the percutaneous resorption of a number of essential oils and found that after 0.5-2 h all essential oils investigated were detected in the expiration air.

Later, other workers stated a good resorption of essential oils through the skin, and data were published on the resorption dependant on time. Katz (2) demonstrated that turpentine oil, 1.8-cineole and α-pinene were resorbed within 20 min, eucalyptus oil and thyme oil, eugenol, linalool, anethole, linalylacetate, geranyl acetate and methylnonylketone within 20-40 min, anise oil, bergamot oil, lemon oil and methyl salicylate within 40-60 min, citronella oil, pine needle oil, lavender oil, geranium oil and cinnamaldehyde within 60-80 min and coriander oil, common rue oil, peppermint oil, citral, citronella oil and geraniol within 100-120 min. Lately Römmelt et al. (3) came to the same results and stated that the skin penetration of terpenes, when applied in a bath, is 100 times better than that of water and 10 000 times better than that of sodium chloride. It was also proved that terpenes were constantly resorbed and eliminated by the blood-vessels, so that a flowing balance occurred. The changes of the concentration in the blood and the expiration air of α- and β-pinene, camphene, limonene, camphor and borneol were described (4). The tests were performed with pine needle and rosemary oil.

Hyperaemic effect

Essential oils within a hyperaemic effect do not only cause a better local blood-circulation, which can easily be observed by a redness of the skin, but they are also influencing the inner organs. The local skin-irritation, the 'primary irritation', sets free mediators in the body (e.g. bradykinin), which in turn cause vasodilatation. The local skin-irritation can also have an influence on the inner organs (relief of pain, effect of 'cardiac ointments' in some cases of angina pectoris). The primary irritation also causes humoral reactions, the results of which can be observed as an antiinflammatory effect. Thus, besides an agreeable feeling of warmth, relief of pain (e.g. muscle and nerve diseases) also antiinflammatory effects can be observed.

Dependant on the degree of skin-irritation one distinguishes between rubefacients, vesicants and suppurants. The rubefacients cause only an increased blood-circulation. It can be observed as redness of the skin. Rubefacients are often used in therapy. One distinguishes between external antirheumatics and antiinflammatory agents. The indications include rheumatic diseases of the joints, lumbago, neuritis, sciatica, shoulder-arm syndroms, etc. as well as bloodless sport injuries such as sprain, bruise, crush, etc.

In rubefacients one usually finds some of the following essential oils or compounds isolated from them: eucalyptus oil, gaultheria oil, juniper berry oil, rosemary oil, rectified turpentine oil, camphor and methyl salicylate. Pharmaceutical preparations are liniments, tinctures or alcoholic solutions, extracts and oils for bath, ointments and gels.

The vesicants cause far stronger skin-irritation and may even cause blisters. They are sometimes used in the treatment of some chronical diseases. The essential oils mentioned above will only cause blisters, when applied in relatively high doses over a longer period of time. Typical vesicants are cantharidine, mustard oil, capsaicine as well as croton oil, the last of which should not be used at all.

The suppurants cause gleety blisters. However, suppurants are obsolete since they may cause necrosis.

Antiinflammatory effect

There are many reasons for the antiinflammatory effect of essential oils, and it varies for different oils. One part of the reaction mechanism may be the starting of secondary biochemical processes, whereby humoral mechanisms stimulate a number of leucocytary reactions (5). Although a series of tests on animals have been carried out (6-8), the mechanisms have so far not been fully explained. However, one seems to agree on that essential oils through skin-irritation are setting free or are binding endogenous substances, whereby the local inflammation process is influenced.

The antimicrobial action of essential oils is also an important aspect of their antiinflammatory effect. Also the skin irritating effect which causes a better local blood-circulation is of importance. Often local inflammations start because of disturbances in the blood-circulation, caused by contraction of the arteroles.

An antiinflammatory effect has been proved for the following plants and their essential oils: chamomile containing matricine, chamazulene and (-)-α-bisabolol (9), yarrow flower, arnica flower containing thymol in the essential oil as well as the sesquiterpene lactones helenaline and helenaline acetate in the tincture (10), and turpentine oil.

The pharmaceutical antiinflammatory preparations are mostly alcoholic extracts (tinctures), ointments and gels in which alcoholic extracts are incorporated.

In some commercial chamomile preparations it has been found that semisynthetic guaiazulene has been added in order to obtain the azulene content claimed for the preparations (11). However, guaiazulene has a markedly less antiinflammatory effect than the naturally occurring matricine from chamomile. This holds also true for synthetic bisabolol and (+)-α-bisabolol, which is isolated from *Populus balsamifera*. Also these bisabolols are used in order to give the preparation the 'chamomile-bisabolol'-content claimed.

Differences in therapeutical activity of chamomile preparations may be found in adulterations of the kind mentioned.

Antiseptic effect

The antiseptic activity of many essential oils has been known for a long time and many microbiological studies have been performed on essential oils (12). However, little is known about the mechanism of action of essential oils. Their lipoid solubility and, therefore, their possibility to penetrate into the cells may give rise to influence on the metabolism of the micro-organisms and thus give an explanation of the effect. This is in agreement with the observation of Malowan (13), who found that the antiseptic activity of many compounds is dependant on their lipoid solubility.

The bacterostatic activity of essential oils can partly be explained by their surface activity (14).

The phenol-coefficient of Rideal et al. (15) and Martindale (16) shows that various essential oils have a highly different bactericide effect. The phenol-coefficient indicates how many times weaker or stronger a bactericide effect of an essential oil is when compared with phenol (= 1.0). Table II shows the phenol-coefficients of some essential oils and some compounds isolated from such oils.

Table II. Phenol-coefficients of some essential oils and pure compounds isolated from essential oils.

Essential oil	Compound	Phenol-coefficient
Anise seed oil		0.4
Peppermint oil		0.7
	Menthol	0.9
Lavender oil		1.6
Lemon oil (Java)		2.2
	Cinnamaldehyde	3.0
	Citral	5.2
	Camphor	6.2
Clove oil		8.0
	Eugenol	8.6
Fennel oil		13.0
Thyme oil		13.2
	Thymol	20.0
	Chlorothymol (synthetic)	75.0

The phenol-coefficient test gives only a relatively rough impression of the bacteriostatic or bactericide activity. The method has been critically investigated later on (17).

In a number of papers (18) the fungistatic and fungicide activity of essential oils have been studied, such as fennel oil, clove oil, cinnamom oil and thymol. They were active against *Candida albicans*, *Sporotrichon* and *Trichophyton* species.

Studies have also been performed on the fungistatic, respectively fungicide activity of chamomile oil, chamazulene, (-)-α-bisabolol, etc. (19,20). Recently also the fungicide effect of chamazulene and (-)-α-bisabolol on *Trichophyton rubrum*, *T. mentagrophytes*, *T. tonsurans*, *T. quinckeanum* and *Microsporum canis* was investigated (21). Both compounds had a fungicide activity in relatively high doses, (-)-α-bisabolol showed a stronger activity than chamazulene. In concentrations of 200 μg/ml both compounds were active. Similar results were reported recently (22).

Granulation stimulating effect

The granulation stimulating effect is usually coupled with an anti-inflammatory and a slightly hyperaemic effect. It is, however, restricted to a few essential oils, such as chamomile oil and preparations containing this oil. The preparations are applied as baths as well as alcoholic chamomile extracts which are applied either directly (in diluted form) or in ointments or gels.

Tests with Japanese rectified peppermint oil on animals with arteficial abrasions or scalds showed a favourable epithel activating activity on abrasions, whereas on scalds no curing effect could be observed (23).

In popular medicine St. John's wort oil (Oleum Hyperici) is applied to heal wounds (24). Oleum Hyperici is an extract prepared by means of olive oil from flowers of *Hypericum perforatum*. This so-called 'Red Oil' contains besides the essential oil also hypericin and pseudohypericin (dianthrones) as active components (25).

Experiments carried out by us with alcoholic extracts and ointment containing such extracts of propolis, the 'lute resin' of bees, showed an antiseptic and a granulation stimulating effect. Alcoholic propolis extracts contain also essential oils.

Deodorizing effect

Best known is the deodorizing effect of chamomile preparations. In surgery after an accident chamomile extracts are used in a bath in order to clean large wounds (26). Also for 'open legs' baths containing chamomile infusions or alcoholic extracts seem to be useful. Deodoration of malodorous wounds are carried out with rectified citrus oils ('terpene free'), rectified turpentine oil, thyme oil and St. John's wort woil.

Insecticide and insect repellent effect

Essential oils have according to literature usually only weak insecticide activity. In 1984 a number of essential oils as well as components from such oils were examined for their insecticide activity (27). Only camphor, citronella oil, eucalyptus oil, clove oil and cinnamom oil as well as citral, carvone and cinnamaldehyde showed such activities. The latter three compounds showed in a 5% concentration a death rate of more than 25%. The essential oil of tetraploide *Calmus* species, the so-called Indian Jammu oil, containing up to 90% of β-asarone has an effect on insects as a chemical sterilizer. Experiments on pupas and adult forms of *Dysdercus koenigii* resulted in an irreversible damage of the ovaries. The effective concentrations were 10-30 μl per 1.5 l of air. The effect of β-asarone is highly specific and a new type of antigonadale agent without side-effects seems to have been found for insect control.

INTERNAL APPLICATION

By inhalation essential oils are usually getting quicker into the blood-circulation than by oral application. On resorption from the stomach and intestines the resorption will be influenced by the food present. Thus, fat containing food will, because of the lipoid solubility of essential oils, have influence on the resorption. In no way the resorption found on oral application can be regarded as a proof for a therapeutic effect, since the elimination speed often is so high that an effective concentration of the essential oils can not be established in the blood or tissue (27). Some compounds of the oils may also undergo chemical changes in the stomach and intestines; so citronellal can be transformed into a cyclic compound. Most of the essential oils resorbed are excreted via the lungs, only a small part via the urine. Differences are, however, known. Borneol is in persons with a healthy liver coupled with glucuronic acid to a degree of 60-100% within 24 h and excreted via the urine. When the excretion takes place via the kidneys, the different compounds of essential oils are usually coupled with glucuronic or sulphuric acid.

Expectorating effect

The expectorating effect is mainly caused by secretolytic and secretomomotoric effects and less by a partial bronchospasmolytic effect. Schäfer and Schäfer (28) showed that on intratracheal application in animals of a mixture of eucalyptus and conifer oil, menthol and camphor a circa 50% reduction of acetylcholine induced bronchospasms was obtained. In clinical studies, however, an obstructive deterioration was observed (29). In tests with animals as well as in clinical tests, the secretolytic and secretomotoric effects of several essential oils were clearly demonstrated. A pharmacological test here is the study of the mucociliaric activity on cilium epithel preparations of frogs. Already 50 years ago the expectorating effect of various essential oils was studied, using cilium epithel of the mucous membrane of the throut and of the oesophagus of frogs (30). In 1980 Müller-Limmroth and Fröhlich investigated the increase as well as the decrease of the mucociliar activity on cilium epithel preparations of frogs. They tested mucilage and saponine-containing crude drugs as well as aqueous extracts of anise seed, fennel and thyme (30). The results showed that anise seed and fennel increased the mucociliar activity (fennel stronger than anise seed), whereas thyme decreased this activity by about

35%. The obstructing effect of thyme and thyme preparations should, however, not cause any doubt about the use of such preparations as effective expectorants, since the test method mentioned above only demonstrates a partial mechanism of the expectorating process, as pointed out by Gordonoff (31), and because the test was performed with aqueous extracts.

After a systemic resorption or after inhalation the essential oils arrive in the bronchi and the main part will be exhaled by the lungs. In both cases, however, they will cause an increased secretion (= a protective reaction) by a direct action on the tracheal and bronchial mucous membrane. In the case of a systemic resorption, a secondary reaction mechanism is possible, i.e. an indirect reflectoric dislocation of an increased bronchial secretion (32). By stimulating the sensible nerves of the mucous membrane of the stomach, an increase of the bronchial secretion reflectorily takes place.

The following essential oils and compounds have expectorating activity without any doubt: anise seed, eucalyptus, fennel, pine needle, dwarf pine, turpentine, thyme and wild thyme (*Thymus serpyllum*) oil as well as camphor and menthol.

In expectorants often also other essential oils are used: chamomile, orange flower, orange peel, peppermint, sage and cinnamom oil.

The oils and isolated compounds of them are applied by inhalation or by aerosol therapy (33) for unspecified affections of the respiration tract and lungs. Also alcoholic solutions, extracts and distillates in syrups, as bath additives, sucklozenges and infusions are used.

A real antitussive activity, as known for codeine, is not known for essential oils. The antitussive effect claimed for some essential oils is probably only a 'positive side-effect'.

Stimulation of the secretion of digestive glands

A number of essential oils are used because of their appetite stimulating, choleretic, cholekinetic and carminative effects - also as spices in food. Some essential oils increase the secretion in the stomach and intestines, other stimulate the liver or gall bladder (32).

Essential oils applied for their effect on the stomach are anise seed, calmus (except the tetraploide), angelica, fennel, orange peel, peppermint and cinnamom oil. Often alcoholic or vinous extracts of the crude drugs are used.

Crude drugs and essential oils with an influence on the gall bladder (cholekinetic effect) are curcuma root oil containing the sesquiterpene phenol xanthorrhizol, and p-tolylmethylcarbinol (an artefact with cholagogic activity, formed during the preparation of the crude drug), calmus, caraway seed, lavender and peppermint oil.

The choleretic effect (increase of the bile secretion) is not so distinct as the cholekinetic effect, but it is mostly coupled with it. Special choleretic effects have been demonstrated for peppermint oil and menthol as well as for curcumin from *Curcuma xanthorrhiza* (32). Cholagogs commonly used contain often borneol, camphor, 1.8-cineole, α- and β-pinene, menthol and menthone.

The carminative effect (on stomach, intestines and gall bladder) of essential oils is mainly due to:

- a local irritating effect of the lining of the stomach, the gastric mucose;
- a reflectoric stimulation of the secretion by the gastric cells and thus improvement of the digestion;
- a spasmolytic effect and relief of flatulence;
- an antiseptic effect (on micro-organisms)
- a cholagogic effect.

Essential oils with carminative effect are anise seed, basil, fennel, chamomile, coriander, caraway seed and peppermint oil; pharmaceutical preparations are alcoholic extracts with or without addition of essential oils, alcoholic distillates and infusions.

Spasmolytic effect

The spasmolytic effect is observed as relief of spasms of smooth muscles, e.g. of intestines and gall bladder as well as of the bronchi. Spasmolytic activity has been found and pharmacologically proved for a number of essential oils, such as chamomile [also (-)-α-bisabolol (34)], caraway seed, fennel, orange and peppermint oil (32,35), as well as balm and cinnamom oil.

Antiinflammatory and antiseptic effects

Matricine in chamomile oil has a strong antiinflammatory activity, as demonstrated in pharmacological tests (9). Preparations containing matricine are used internally for the treatment of gastritis, stomatitis simplex and even stomatitis ulcerosa.

The antiseptic effect is especially important in the treatment of colds. Inhalation and gargling with solutions of sage, eucalyptus, thyme and cinnamom oil, as well as thymol are mostly used. Also for the treatment of inflammations in the oral cavity such oils have successfully been used (36).

Diuretic effect

Essential oils have no real diuretic activity and they are only partly excreted by the kidneys. Freund (37) found that juniper berry extracts had no diuretic activity when the essential oil or the terpene hydrocarbons were removed, whereas Schneider (38) claimed that the 'terpene free' oil containing mainly terpinen-4-ol had a distinct diuretic activity.

The following essential oils are, however, not suitable for treatment of oedema or hypertonia: onosis root, lovage root, orthosiphon leaf, parsley fruit and root, and juniper berry oil.

Sedative effect

As to the claimed sedative effect of many essential oils, little has been proved experimentally except for valerian (39,40). The valepotriates have been shown to be active tranquilizing compounds (41). Whereas *Valeriana officinalis* contains only small amounts of valepotriates (0.8-1.7%) other *Valeriana* species (*V. wallichii* and *V. edulis*) contain 2.8-3.5% and 8-12% respectively. However, it seems that also the essential oil of valerian plays a role for the sedative effect, as already stated by Schmiedeberg (42) and confirmed by Ammon (personal communication).

For balm, hop and St. John's wort oil, all of which are claimed to have some sedative activity, no proofs are available. In animal tests with balm oil a certain suppression of the spontaneous mobility was observed; it held also true for citronellal and citral, the main components of the oil (43). For hop a sedative-hypnotic effect was observed in animal tests (44, 45) and for St. John's wort oil a slight antidepressive effect (46).

Circulation stimulating effect

Some essential oils and compounds occurring in them seem to have a circulation stimulating activity. However, the effect is usually only observed with high doses. A stimulus of the medullary centres and the respiration, heart and circulation can also occur reflectorily. In former times this effect was often used in form of 'smelling salts'.

The value of camphor as a central analeptic is not judged unanimously. Camphor is little used internally because of its many side-effects. However, still some cardiacs contain camphor. They are used for a symptomatic treatment of functional heart trouble in the form of camphor-containing 'heart ointments'.

Camphor has a stimulating effect on the respiration and circulation centres, e.g. by influencing the activity of the vegetative centres, it has indirectly also peripheric effects on heart and blood vessels (47). Oleum camphoratum for injection is still applied by circulatory collapses (Schimmel, personal communication).

Rosemary oil is, because of its content of borneol, bornyl acetate, 1.8-cineole and camphor also used for circulation stimulation. Such an activity is also claimed for lavender and peppermint oil, especially on inhalation.

UNDESIRED SIDE-EFFECTS

Undesired side-effects of essential oils have been observed on misuse of such oils for abortion or on application of such oils in excessive doses. Essential oils with a high percentage of unsaturated compounds are generally more toxic (48,49).

Allergic effect

Contact induced effect is often found for essential oils rich in terpene hydrocarbons (50,51) (non-rectified turpentine oil). However, according to Hausen (52) are not the terpene hydrocarbons causing the allergy, but the hydroperoxides, which are formed on processing and storage.

In most cases the allergene is not known; an exception is, however, cinnamaldehyde in cinnamom oil. Turpentine oil can cause professional dermatitis for painters, but also citrus oil is said to be active. Some citrus oils show also phototoxic and photosensitizing activity because of the furanocoumarins in these oils. The essential oil of *Arnica montana* has in contrast to arnica tincture only little allergenic activity. The risk of allergy when using chamomile preparations can be neglected.

According to Hausen (personal communication) the following essential oil-bearing plants may show some contact-allergic activity: chamomile, lavender, balm, parsley, peppermint and sage.

Phototoxic and photosensitizing effect

Some essential oils containing furanocoumarins can after an initial photosensitizing effect give phototoxic reactions, which can be observed as erythem and formation of blisters. In therapeutic doses and limited application, the furanocoumarins are used in the treatment of vitiligo. After an initial formation of erythem, a distinct pigmentation takes place. Other, simple coumarin derivatives, such as umbelliferon, herniarin and aesculetin have a sunlight filter effect because they absorb UV-light of 280-315 nm.

Phototoxic and/or photosensitizing activity have been found for the following essential oils: citrus oils, especially those produced by extraction, and bergamot oil. These oils are responsible for the redness of the skin which can be observed on application of Eau de Cologne and exposure to sun (Berloque dermatitis, photodermatitis pigmentaria). Some Apiaceae oils (angelica, parsley fruit and root) and particularly that of *Heracleum mantegazzianum* (53) can cause phototoxic reactions.

Necrotic effect

Long term application and/or high doses of the essential oil of *Juniperus sabina* can cause necrosis. Even an external application of the oil can lead to intoxications with sincere damage to the kidneys.

Narcotizing effect

Narcotizing-paralysis has been found only for a few essential oils and in high, toxic doses (54). The reactions are irreversible and seem to be caused by a non-specific damage of the tissue.

The following essential oils and oil components can lead to paralysis on application of high doses: essential oils from *Thuja* species as well as α- and β-thujone (55). Non-rectified turpentine oil and apiole. Vermouth oil leads to psychic and motoric disturbances by damage of the central nervous system, because of its thujone content.

Abortion provoking effect

The uterus stimulating effect of some essential oils has been and is still misused for abortion. Essential oils do not only cause a distinct filling of the blood vessels of the pelvis and a strong contraction of the uterus, but also a degenerative transformation of parenchymatic organs (especially of the liver). Misuse of essential oils may therefore be fatal.

For the following essential oils and components abortion provoking activity has been observed in high doses: thujone-containing essential oils such as *Juniperus sabina* and *Thuja occidentalis* oil, as well as chenopodium oil (*Chenopodium ambrosioides*), apiole and parsley fruit oil from the apiole variety, containing 60-80% apiole. The essential oils and compounds mentioned are active only in high doses.

Nephrotoxic effect

Juniperus berry oil has a distinct kidney irritating/kidney injuring effect because of a high percentage of monoterpene hydrocarbons. The terpene alcohol terpinen-4-ol seems not to have any irritating effect (56). Severe damages of the kidneys could be observed on oral application of turpentine and sandelwood oil. Tukioka (57) described histologically proved damage of the kidney epithel on application of these oils.

Hepatotoxic effect

In addition to thujone also thymol (used as a choleretic) and turpentine oil (used as a diuretic) on oral application in high doses may damage the liver. In animal tests also safrole caused such damage (58).

Cancerogenic effect

Taylor et al. (59) found that calmus oil from the tetraploide Indian Jammu-variety caused tumours in the duodenal ulcer area in rats after about 60 weeks. The food contained 500, 1000, 2500 and 5000 ppm of calmus oil. The Indian calmus oil contains about 80% of β-asarone, the crude drug about 5.5%. The diploide *Acorus calamus* originating from North America contains no β-asarone, the East European calmus (triploide), which is cultivated in Europe, contains less than 10% β-asarone in the oil and maximum 0.3% β-asarone in dried plant material (60).

Carcinogenic activity (liver carcinome) may also be caused by safrole and dihydrosafrole (61). Safrole is the main component of sassafras oil (75-90%) from the root wood of *Sassafras albidum* (= *S. officinale*). In some countries safrole is, therefore, not permitted in food (62).

REFERENCES

(1) Strähli W. 1940, Thesis, University of Bern, Switzerland.
(2) Katz AE. 1947, Parfüm. mod. 39:64.
(3) Römmelt H, Zuber A, Dirnagl K, Drexel H. 1974, Münch. Med. Wochenschr. 116:537.
(4) Römmelt H, Drexel H, Dirnagl K. 1978, Heilkunst 91,No.5:21.
(5) Gildemeister E, Hoffmann Fr. 1956, Die ätherischen Öle, Band I, Berlin, Akademie-Verlag, p. 112.
(6) Laden C, Blackwell RQ, Fosdick L. 1958, Am. J. Physiol. 195:712.
(7) Atkinson D, Hicks R. 1975, Agents Actions 5:239.
(8) Bonta I, Nordhoek J. 1973, Agents Actions 3:348.
(9) Jakovlev V, Isaac O, Flaskamp E. 1983, Planta Med. 49:67.
(10) Willuhn G. 1981, Pharm. unserer Zeit 10,No.1:1.
(11) Messerschmidt W. 1973, Dtsch. Apoth.-Ztg. 113:745.
Herrmann R. 1982, Dtsch. Apoth.-Ztg. 122:1797.
(12) See (5), p. 124.
Kienholz M. 1959, Arzneim.-Forsch. 9:519.
Ritzerfeld W. 1959, Arzneim.-Forsch. 9:521.
(13) Malowan SL. 1931, Z. Hygiene 1,No.1:93.
(14) Rideal S, Rideal EK, Scive A. 1928, Parfum. Record 19:285.
(15) Rideal EK, et al. 1930, Parfum. Record 21:344.
(16) Martindale WH. 1910, Parfum Record 1:266.
(17) Borneff J, Pfeiffer E. 1971, Gutachten über die Testung des Japanischen Heilpflanzenöles (IHP-Rödler) auf Bakterizide.
(18) See (5), p. 140.
(19) Szalontai M, Verzár-Petri G, Florian E, Gimpel F. 1975, Pharm. Ztg. 120:982 and 1975, Dtsch. Apoth.-Ztg. 115:912.
(20) Szalontai M, Verzár-Petri G, Florian E. 1977, Parfüm. Kosmet. 58:121 and 1976, Acta Pharm. Hung. 46:232.
(21) Schilcher H, Beyerle-Müller HL. Eur. J. Appl. Microbiol. Biotechnol., in press.
(22) Janssen AM, Scheffer JJC, Baerheim Svendsen A, Aynehchi Y. 1984, Pharm. Weekbl. Sci. Ed. 6:157.
(23) Borneff J, Graf Z. 1971, Gutachten über die Testung der Wirkung des Japanischen Heilpflanzenöles (IHP-Rödler) auf die Wundbehandlung.
(24) Weiss RF. 1982, Lehrbuch der Phytotherapie, 5. Auflage, Stuttgart, Hippokrates-Verlag, p. 379.
(25) Schilling W. 1969, Präp. Pharm. 5:125.
(26) Hänsel R, Haas H. 1983, Therapie mit Phytopharmaka, Berlin/Heidelberg/New York/Tokyo, Springer-Verlag, p. 270.
(27) See (5), p. 119.
(28) Schäfer D, Schäfer W. 1981, Arzneim.-Forsch. 31:82.
(29) See (26), p. 113.
(30) Müller-Limmroth W, Fröhlich HH. 1980, Fortschr. Med. 98:95.
(31) Gordonoff T. 1938, Ergeb. Physiol. 40:53.
(32) See (5), pp. 195, 200, 210.
(33) Wilde W. 1960, Prakt. Arzt, Wien 14:278.
Böhlau V, Böhlau E. 1958, Die Inhalationsbehandlung mit Aerosolen, Leipzig, VEB Georg Thieme-Verlag, p. 3.
(34) Achterrath-Tuckermann V, et al. 1980, Planta Med. 39:38.
(35) Forster HB. 1979, Thesis, University of Tübingen, FRG.
(36) Hunkirchen M. 1970, Dtsch. Zahnärztebl. 24,No.3:8.
(37) Freund G. 1937, Thesis, University of Bern, Switzerland.
(38) Schneider G. 1975, Pharmazeutische Biologie, Mannheim/Wien/Zürich, Wissenschaftsverlag, p. 128.

(39) Becker H, Reichling J. 1981, Dtsch. Apoth.-Ztg. 121:1185.
(40) Becker H. 1983, Dtsch. Apoth.-Ztg. 123:2470.
(41) Thies P. 1967, Dtsch. Apoth.-Ztg. 107:1411.
(42) Schmiedeberg O. 1913, Grundriss der Pharmakologie, 7. Auflage, Leipzig, L.W. Vogel.
(43) Wagner H, Sprinkmeyer L. 1973, Dtsch. Apoth.-Ztg. 113:1159.
(44) Hänsel R, Wohlfart R, Schmidt H. 1982, Planta Med. 45:224.
(45) Wohlfart R. 1983, Dtsch. Apoth.-Ztg. 123:1637.
(46) See (24), p. 332.
(47) See (26), pp. 49, 225.
(48) Macht H. 1938, Am. Med. Assoc. 110,No.6:408.
(49) Macht H. 1939, Arch. Int. Pharmacodyn. 63:43.
(50) Urbach E, Wiethe C. 1932, Naunyn-Schmiedebergs Arch. Exp. Pathol. Pharmakol. 281:253.
(51) Wyss-Chodat F. 1939, Schweiz. Med. Wochenschr. 105.
(52) Hausen BM. Plant Dermatitis - A manual of Sensitizing Species, Berlin/New York, Walter de Gruyter, in preparation.
(53) Frohne D, Pfänder HJ. 1982, Giftpflanzen, Stuttgart, Wissenschaftliche Verlagsgesellschaft, p. 46.
(54) See (5), p. 113.
(55) Wenzel DG, Ross CR. 1957, J. Am. Pharm. Assoc. 46:77.
(56) See (26), p. 209.
(57) Tukioka M. 1927, Proc. Imp. Acad. (Tokyo) 3:624.
(58) Anonymous (FDA). 1957, Drug Cosmet. Ind. 81:592.
(59) Taylor JM, et al. 1967, Toxicol. Appl. Pharmacol. 10:405.
(60) Stahl E, Keller K. 1981, Pharmazie 36:53.
(61) Gildemeister E, Hoffmann Fr. 1966, Die ätherischen Öle, Band IIId, Berlin, Akademie-Verlag, p. 454.
(62) Walker GT. 1968, Seifen Öle Fette Wachse 94:451.

Note from the Editors: more details can be found in the paper entitled 'Ätherische Öle - Wirkungen und Nebenwirkungen' by Schilcher H. 1984, Dtsch. Apoth.-Ztg. 124:1433.

ARZNEIPFLANZEN IM SPIEGEL NIEDERLÄNDISCHER MALEREI*

F.-C. CZYGAN

ABSTRACT

Medicinal plants and their compounds (such as essential oils) have been presented in the fine arts of many periods of civilization. They have been from former times up to now not only subjects of lyric poetry and prose, but also of pictorial art. In paintings medicinal plants – just as many other plants – were not used only because of their decorative characters, but also for their symbolical significance. This symbolism can be demonstrated by many examples. It is especially true as far as the symbolical character of certain medicinal plants within Dutch painting is concerned.

May I, as an amateur in the history of art, be allowed to extend the conception of Dutch painting in a certain European sense, for it was the Dutch painters who – in spite of typical Dutch subject and details in their paintings – have never been restricted to the region of the historical Netherlands.

The red thread of this lecture will be similar to a meander. I will take your ears and eyes and lead you through my artificial garden with blooming plants and ripe fruits. During this walk you will meet for example *Jan Breughel's* famous picture '*Flower bunch in a wooden pot*', to-day in the Alte Pinakothek of Munich. There is *Frittilaria imperialis* in the center of the bouquet, surrounded by *Lilium candidum*, *Lilium martagon*, *Iris species*, and *Lilium bulbiferum* and *roses*, all old symbols, already mentioned in the middle-ages and referring to the virginity and innocence of St. Mary. I will take you to the *Portinari* altar-piece, painted by *Hugo van der Goes* in Florence, with *Iris*, *Dianthus*, *Aquilegia* and *Lilium bulbiferum* in front of the crib. I will try to explain the symbolism of the pomegranate used already in art by Old Egyptians, but also by *Abraham Breughel* in 1669 or by the modern artist *E. Schlotter* in his picture '*Nocturno*', which dates from 1977. This fruit full of seeds, with its sweet and sour taste symbolizes not only life, fertility and productivity, eros and voluptuousness, but also charity and the wisdom of old age. With these and other examples I will demonstrate to you both the beauty of Dutch painting and the plant symbolism implied, which had such a long history from ancient periods until to-day.

Arzneipflanzen und ihre Inhaltsstoffe, z.B. ätherische Öle, charakterisiert durch den Geruch oder Duft, sind in vielen Kulturen in künstlerischer Weise dargestellt worden (1,2). Bis weit in die Zeit der ägyptischen Hochkulturen lassen sich Beispiele zurückverfolgen. Medizinalpflanzen waren und sind

*Meiner Mutter in Dankbarkeit gewidmet

A. Baerheim Svendsen and J.J.C. Scheffer (eds), Essential oils and aromatic plants, ISBN-13: 978-94-009-5137-2

nicht nur Objekte der Malerei und Plastik – erinnert sei an die Blattmotive der Kapitelle romanischer und gotischer Kirchen des Mittelalters (3) – , sondern auch der Poesie und Prosa (4). Eine wichtige Rolle spielen leicht flüchtige Naturstoffe in der Literatur. Marcel Prousts '*Suche nach der verlorenen Zeit*' wäre ohne den Duft, der ihm längst vergangene Tage in die Erinnerung zurückrief, nicht denkbar (4).

In der Malerei der Alten wurden Pflanzen, vor allem solche, mit denen der Mensch täglich in Berührung kam, seien es Nahrungs- oder Heilpflanzen, aber auch rituelle Zauberpflanzen, nicht nur ihres dekorativen Charakters wegen, sondern auch ihres Symbolwertes wegen dargestellt. Diese Typisierung von Pflanzen und ihre Zuordnung zu bestimmten legendären Personen oder zur geschichtlichen und religiösen Gestalten und Ereignissen, ihre Einordnung in besondere Charakterbereiche, muss natürlich verstanden werden aus der Mentalität der Betrachter, für die sie geschaffen wurden. Nicht *art pour art* stand hier im Vordergrund des Schaffens, die Information der des Schreibens und Lesens nicht Kundigen war die Aufgabe der Symbole. Mit dieser Aufgabe stimmt überein, dass von der ägyptischen Zeit über die Antike bis zum Mittelalter in vielen Fällen der Künstler hinter sein Werk zurücktrat und unbekannt blieb. Erst in der Renaissance, die das Individuum in den Mittelpunkt rückte, trat der 'Produzent' gleichwertig neben sein Oeuvre.

Diese Charakterisierung von Pflanzen und ihrer Symbolkraft (5-7) findet auch in der niederländischen Malerei ihren Niederschlag. Es sei mir – als kunsthistorischem Laien – erlaubt, den Begriff der niederländischen Malerei weit zu fassen und nicht geographisch stark einzuengen. Gerade die niederländische Kunst ist trotz vielseitig genutzter holländischer Sujets und Details immer europäisch gewesen. Mein kunsthistorischer Spaziergang durch die Welt der holländischen Gemälde, die Arzneipflanzen enthalten, wird das zeigen. Diese Exkursion ist geprägt von persönlichen Vorlieben und sicherlich auch von Vorurteilen. Sie ist nicht ausgewogen und ein Ausflug sui generis. Einem Mäander gleich soll der Rote Faden sein, der uns führt. Es soll versucht werden, die Schönheit der Bilder nahe zu bringen, die Symbole von obsoleten oder heute noch aktuellen Arzneipflanzen und ihre Bedeutung anhand von Gemälden des 15. Jahrhunderts bis in unsere Zeit zu erklären.

Mir ist klar, dass – definiert man den Begriff '*niederländische Malerei*' im heutigen kunstgeschichtlichen Verständnis – damit die Kunst der Niederlande und Belgiens gemeint ist. Hier lagen die künstlerischen Blütezeiten in der Spätgotik und im Barock. Sicherlich gehören die Werke von Hubert (ca.

1366-1426) und Jan (ca. 1390-1441) van Eyck aus dem niederländisch-burgundischen Umkreis, z.B. der '*Genter Altar*', zu den Meisterleistungen abendländischer Kunst und Kultur. Im gleichen Atemzug sind aber auch zu nennen Rogier van der Weyden (1399/1400-1464) und sein Lehrer, der Meister von Flémalle, Hans Memling (1433-1494), Hugo van der Goes (1440/45-1482), Dieric Bouts (ca. 1415-1475) und später Hieronymus Bosch (1453-1516) mit seinen berühmten Höllenvisionen. Die Renaissance im 16. Jahrhundert ist für uns mit dem Namen Pieter Breughel d.Ä. (1528/30-1569) verbunden. Maler wie Barend van Orley (ca. 1488-1542), einer der nordischen Manieristen und Romanisten, Lucas van Leyden (1489 o. 1494-1533) u.a.m. standen ihm zur Seite. Der Barock des 17. Jahrhundert brachte vor allem im Hochbarock eine deutliche Trennung in die höfische Form der spanisch-katholischen südlichen und der mehr bürgerlichen Form der nördlichen Niederlande. Der flämische Barock (Namen wie Anton van Duyck, 1599-1641, Jakob Jordaens, 1593-1678, Frans Snyders, 1579-1657, seien genannt) wurde besonders von Peter Paul Rubens (1577-1640) geprägt. Der protestantische Barock war in seinen Grundelementen realistischer, was sich deutlich in den aus dieser Zeit bekannten Bildgattungen zeigt. Gewisse Spezialisierungen waren zu erkennen: u.a. Gruppenbilder von Frans Hals (1581/85-1666), Stilleben von Willem Kalf (1619-1693), Pieter Claesz (1597/98-1661), Genrebilder von Gerard Terborch (1617-1681), Jan Steen (1625/-1679), Jan Vermeer (1632-1675), Landschaftsbilder von Jan Josephsz. van Goyen (1596-1656), Seestücke von Willem van de Velde (1633-1707), Portraits u.v.a.m. Über allen Typisierungen dieser Art stand das Werk des Leidener Müllersohns Rembrandt Harmenesz. van Rijn (1606-1669).

Zwei Jahrhunderte später, im 19. und 20. Jahrhundert waren es u.a. Malernamen wie Vincent van Gogh (1853-1890), Theo van Doesburg (1883-1931), Piet Mondriaan (1872-1944), Jan Sluijters (1881-1957), Frans Masereel (1889-1972) und René Magritte (1898-1967),die der belgischen und niederländischen Malerei Weltruhm verschafften.

Lassen Sie mich mit einem niederdeutischen Meisterstück, das im weitesten Sinne dem Thema unterzuordnen ist, beginnen, mit einem Werk des '*Meisters des Göttinger Barfüsser-Altars*', das um 1425 entstanden ist (I)*. Hier haben wir eine besondere Form von Arzneipflanzengarten mit einer Vielzahl von Heilpflanzen: den 'Beschlossenen Garten'. Entsprechend dem Hohelied Salomos

---

* Die römischen Zahlen geben die in der Totale und im Detail während des Vortrags aufgezeigten Gemälde an (s. Seite 246)

(Hohelied 4,12) ist in der Kunstgeschichte der *Hortus conclusus* ein von einem Flechtzaun oder Mauerwerk umgebener Garten, in dem meist Maria alleine oder mit dem Kinde sitzt, oft umgeben von Heiligen und Engeln, immer aber umrahmt von einer Blumenfülle. Der *Hortus conclusus* stellt die Jungfräulichkeit, die unbefleckte Empfängnis Mariens dar. Dazu passen — neben anderen Mariensinnbildern der Keuschheit wie dem versiegelten Brunnen, dem Turm aus Elfenbein — die vielen in ihrer Symbolik bedeutsamen Marien- und Christuspflanzen, die in ihrer Fülle von Lottlisa Behling in dem Buch '*Die Pflanze der mittelalterlichen Tafelmalerei*' 1967 untersucht wurden. Auf diesem Bild ist in den *Hortus conclusus* eine '*Noli-me-tangere*'-Szene hineinkomponiert. Maria Magdalena, vermutlich allgemein das Weibliche darstellend, begegnet dem als Gärtner erscheinenden Christus nach dessen Auferstehung am Ostermorgen. Sie will seine Hände erfassen. Mit den Worten '*Noli-me-tangere*' weist er sie ab (Joh. 20, 14-18). Hier im Gartenmilieu war dem Meister Gelegenheit gegeben, im wahrsten Sinne des Wortes mit vollen Händen Pflanzen in den Garten zu setzen. Nicht willkürlich haben sie ihren Platz. Ganz bestimmte Positionen sind für sie ausgesucht, eben um im Sinne mittelalterliche Symbolik bestimmte Aussagen zu machen.

Von der Fülle der Pflanzen, die diese Begegnung umrahmen, steht zentral ein Strauss mit fünf gelben Narzissen, genau zwischen den Händen der sich Begegnenden. Osterglocke (*Narcissus pseuonarcissus*) heisst diese alte Blume der Passion und der Auferstehung. Auf manchem Bild, das das österliche Geschehen darstellt, ist sie zu finden. Meist in der Fünferzahl, entsprechend den fünf Wundmalen Christi im Kreuz bedeutet die Osterglocke den Sieg über den Tod.

Hinter der Maria Magdalena finden wir die Madonnenlilie (*Lilium candidum*) als ein Symbol der Reinheit, nicht nur der Mutter Jesu, sondern ganz allgemein der weiblichen Komponente im Heilsgeschehen. Hier symbolisiert die uralte Heilpflanze Lilie in besonderer Weise, dass Heil und Heilung nicht nur etymologisch, sondern auch vom ethischen Wert her eng zusammengehören oder zusammengehören mussten. Es bestand zur Zeit der Liliendarstellungen, oft mehr als in der Gegenwart, das Wissen um den Dualismus des Menschen. Nur die Einbeziehung dieser Ambivalenz von Psyche und Physis in der Therapie lassen auch heute noch einen Mediziner zum Arzt werden, der nicht die Krankheit, sondern den Kranken heilt. Heil und Heilung, dieses Spannungsfeld, auf dem sich naturwissenschaftliche und geisteswissenschaftliche Überlegungen treffen, ist auch in der Symbolik der übrigen Pflanzen des '*Noli-me-tangere*'-Bildes zu finden. Ganz nahe Jesu wächst die Stockrose (*Althaea rosea*). Schon ihr

Name (*'altheo'*-ich heile) weist auf die hohe Verehrung dieser Arzneipflanze hin, die zum einen das Leben symbolisiert, zum anderen mit ihrer betonten Pentamerie der Blütenkrone wohl an die fünf Stigmata Christi am Kreuz erinnert.

Zentral hinter der linken Schulter Christi, seinem Herzen nahe, hat der Meister die Akelei (*Aquilegia vulgaris*) hingemalt. Dies betont ihre Bedeutung im Heilsgeschehen. Über Jahrhunderte hinweg hat die bizarre Blüte dieser Pflanze auf fromme oder unfromme Maler eine eigenartige Faszination ausgeübt. Schon bei den Germanen war die Akelei der Göttermutter Frigga und der Freya geweiht. Als marianisches Attribut wurde sie auf die Mutter Christi übertragen. Die Akelei steht aber auch in der Gestalt der Blütenhonigblätter langhalsigen Tauben ähnelnd (engl. *Columbine*-Flower; in manchen Gegenden Deutschlands '*Täuberl*' genannt) für das Tiersymbol der Taube, die ihrerseits die Gaben des Heiligen Geistes symbolisiert. Nach R. Fritz entspricht die Akelei ferner einer auf Amuletten noch in der Rembrandt-Zeit gebräuchlichen Lobpreisung Gottes, der Kabbala-Formel '*AGLA*' = '*Atha Gibbor Leolam Adonai* = Du Held in Ewigkeit, Mein Herr' (Psalm 89,53), die volksetymologisch mit dem mittelhochdeutschen '*aglai*' für Akelei gleichgesetzt wurde (7a).

Auf viele Pflanzen dieses '*Noli-me-tangere*-Bildes' wäre noch hinzuweisen, so auf die Erdbeeren (*Fragaria vesca*), die oft als Speise der Engel gedeutet werden. Früher in der Volksmedizin bei Gemütskrankheiten eingesetzt, sollen sie die bedrückten Sinne Jesu geheilt haben. Demütig beugt sich die Erdbeerstaude, dem wohlriechenden Veilchen (*Viola odorata*) gleichend, vor Christus zur Erde, Bescheidenheit darstellend. Die Trimerie des Blattes erinnert an die Dreifaltigkeit, die blutroten Früchte an das Blut Christi, die pentameren Kronblätter an die fünf Wundmale seiner Passion.

Und schliesslich sei noch eine zur Zeit als Ausgangsmaterial für den wichtigen Arzneistoff Vincamin genutzte Pflanze genannt: das Immergrün (*Vinca minor*). Als Sinnbild der Treue und Beständigkeit war sie, auch Sinngrün genannt, Jungfrauen geweiht und gerade in der mittelalterlichen Tafelmalerei oft dargestellt, so z.B. im berühmten '*Paradiesgärtlein*' von ca. 1440 eines oberrheinischen Meisters. Noch heute finden wir *Vinca minor* in manchen Gegenden als Friedhofspflanze auf Gräbern vornehmlich junger Menschen. '*Mädchenkraut*, *Maidlein*, oder *Jungfernpalm*, *Totenviole*' nennt sie der Volksmund. Und die auch im Winter grünen Triebe von *Vinca* galten als Zeichen der Unvergänglichkeit und Unsterblichkeit. Darüber hinaus gehört *Vinca minor* zu den apotropäischen, d.h. zauberabwehrenden Pflanzen. Schon im *Gart der Gesundheit* von 1485 heisst es: '*Welcher diss krut by ime draget, vber den hait der tüfel kein gewalt*'.

Aus dem *Hortus conclusus* des Mittelalters ist ganz ohne Zweifel der Bildtypus '*Maria im Rosenhag*' hervorgegangen. In vielfältiger Weise hat man hier marianische Blumensinnbilder benutzt. An der Spitze steht die Rose (*Rosa* species), die Königin der Blumen und wohl ältestes Sinnbild Mariens. Sie ist, wie so häufig Blumensymbole, in ihrer Bedeutung ambivalent. So weist die rote Rose auf die Liebe und das Martyrium hin, die weisse Rose auf die jungfräuliche Reinheit Mariens: '..., *die zweierhande varbe hat, rot und wiz. Bi den wizen ir reine magetum, bi den roten ir vollenkumene mine*' (8). Die Fünfzahl mit der Rose verbunden als fünfblütiger Rosenzweig oder als Blüte mit fünf Kronblättern, bedeutet darüber hinaus Verschwiegenheit (erinnert sei an die pentamere Rose im Nimbus über den Beichtstühlen der katholischen Kirche) und gibt einen Hinweis auf die fünf Stigmata Christi am Kreuz. Oft werden die Rosen stachellos dargestellt, besonders in den Händen von Engeln sind sie ein Signum für das Paradies. Erst nach dem Sündenfall wuchsen der Legende nach den Rosen Stacheln. Verschiedentlich wird auch Darstellungen Mariens die dornenlose Rose zugefügt, um ihre Freiheit von der Erbsünde zu symbolisieren (so bei der '*Stuppacher Madonna*' M. Grünewalds). Diese dornenlose 'Rose' ersetzt oft die Pfingstrose oder Paeonie (*Paeonia officinalis*). Sie wird von Konrad von Würzburg in seinem Marienlied als Rose '*an allen stift*' (ohn' jeden Dorn) besungen (9).

Martin Schongauer (1450-1491) hat mit seiner '*Maria im Rosenhag*' von 1478 (II) in besonderer Weise der Marienverehrung Ausdruck verliehen, in seiner Empfindsamkeit wohl nur noch von der feierlichen Idealität der Maria (III) Stephan Lochners (1400/15-1451/52) erreicht. Schongauers enge Beziehungen, die er in der Zeit seiner Wanderjahre durch Burgund, Spanien und den damaligen Niederlanden zu den grossen Holländern Dieric Bouts, Hugo van der Goes und in besonderem zu Rogier van der Weyden hatte, prägten seinen Stil. Manches Bild von Schongauer kann dessen längeren Aufenthalt in der Werkstatt van der Weydens nicht verleugnen. Niederländischer Realitätssinn und die Eleganz Burgunds sind bei Schongauer mit spätgotischer Tradition verknüpft. Neben der Rose und den weiteren Pflanzen dieses Gartens sui generis (z.B. Erdbeeren, die Nachtviole (*Hesperis matronalis*) und eben auch die 'stachellose' Pfingstrose) sind es die schwebenden Engel, die hier die Himmelskönigin krönen wollen und diesem *Hortus conclusus* ihren himmlischen Abschluss geben. Und anders als im Garten des '*Meisters des Barfüsser-Altars*' wird hier das neue Verhältnis zur Natur im 15. Jahrhundert deutlich. Die Natürlichkeit der

Darstellung zeigt das Bemühen, ein Abbild der Natur zu geben. Dieser Schritt, im Bild die reale Umwelt festzuhalten, war selbstverständlich bereits in früheren Jahrhunderten zu erkennen, nicht zuletzt vorbereitet und eingeleitet durch den Realismus in der Philosophie der Schriften der Kirchenväter wie Thomas von Aquino und Albertus Magnus. Sie haben 'durch die Bestimmung der Symbolbedeutung von Pflanzen deren natürliche Darstellung unverwechselbar nach Gattung, Art und Unterschied erst möglich gemacht und ihnen damit den Weg in die realistische und symbolfreie Bilddarstellung geebnet' (10).

Der Umbruch vom Mittelalter zur Renaissance ist ebenso im *Portinaria Altar* (IV) von Hugo van der Goes (1440/45-1482) zu erkennen. Eine Epoche löst die andere ab. Dieser Weg in den Humanismus als Gegenbewegung zur Scholastik und kirchlichen Autorität bedeutete auch für die Malerei Betonung der Realität und des Individuums. 'Die Entdeckung dieser Welt' war nach Kaufmann (11) 'eine unermessliche Erweiterung menschlichen Bewusstseins, Denkens und Fühlens, die nicht abstrakter und bezugsloser Bildungsinhalt war als den wir sie heute zumeist sehen, sondern etwas mit Geist und Sinnen Erlebtes'. Natürlich lebte auch, bewusst oder unbewusst, der Symbolcharakter der Pflanzen weiter.

Die Neigung Hugo van der Goes zum Monumentalen, sein Sinn für das Bühnenhafte, sein darstellerischer Realismus wird im Triptychon des *Portinari-Altars* (um 1473-1475 geschaffen) besonders deutlich. Geöffnet zeigt die Mitteltafel die Anbetung der Hirten. In diese seltsam leere Mitte mit dem strahlenden Licht ist das Jesuskind gebettet, umgeben von der Heiligen Familie, den rauhen Hirten und dem prächtig anzusehenden Engeln. Im Vordergrund steht ein Albarello, ein Marjolikagefäss mit weissen und blauen Iris (*Iris species*) und einer Feuerlilie (*Lilium bulbiferum*), daneben ein venezianisches Rippenglas mit sieben voll erblühten Akelei-blüten und roten Nelken (*Dianthus caryophyllus*). Davor liegen duftende Veilchen (*Viola odorata*). Alle diese Pflanzen haben die zum Teil schon bekannte hohe symbolische Bedeutung. Neben der Feuerlilie weisst auch die Nelke auf die Passion Christi hin. Von besonderem Interesse ist die Zahl 'sieben' der Blüten der Akelei. Wie schon erwähnt, steht sie symbolhaft für die Geisttauben. Ursprünglich betrug die Zahl der Geistgaben sechs: Bei Jesajas (11,12) ist zu lesen: 'Und es wird eine Rute aufgehen von dem Stamm Isais und ein Zweig aus seiner Wurzel Frucht bringen, auf welchem wird ruhen der *Geist des Herrn*, der *Geist der Weisheit und des Verstandes*, der *Geist des Rates und der Stärke*, der *Geist der Erkenntnis und der Frucht des Herrn*'. Im Hohen Mittelalter wurde die Sechszahl auf die heilige Zahl Sieben erhöht: Die *Frucht des Herrn* unterteilte man in *Frömmigkeit* und *Gottesfurcht*. An die bäuerliche Herkunft und Verbindung des Flamen van der Goes erinnert die Getreidegarbe im Vordergrund; vielleicht ist sie auch ein Sinnbild der Fülle des durch die Geburt Christi in die Welt getragenen Neuen Bundes zwischen Gott und den Menschen. Oder hat van der Goes hier an Maria als '*guten Acker welcher ohne Saat Getreidehaufen hervorbrachte und ... eine reiche Ernte feierte*' erinnern wollen. So wird sie bereits im 6. Jahrhundert von Jakob von Batnae in einen Gedicht besungen (nach 13).

In dem erfolgreichen Bemühen der Renaissance, den Menschen nicht nur in seiner morphologischen, sondern auch in seiner sinnlichen Individualität zu sehen, erkenne ich ein Charakteristikum dieses Zeitalters: der Mensch als Sinnen, und damit als Naturwesen. Malereien, die das ausdrücken 'sind ohne die im 16. Jahrhundert vordringende Naturforschung nicht zu denken. Wie die Bildkunst des 16. Jahrhunderts mit Veranschaulichungen der Elemente und der Gezeiten auf die Kosmobiologie eingegangen ist, so hat die Anthropologie in den Fünf-Sinne-Figuren einen Niederschlag gefunden' (11).

Das gilt auch für die allegorischen Darstellungen der '*Fünf Sinne*', die Jan Breughel d.Ä. 1617-1618 gemalt hat. Kauffmann (11) sagt zu diesem Zyklus: 'Da nun in allen fünf Bildern Breughels jeder der Sinne mit gleicher Ausführlichkeit ausgerüstet wurde, bekommen wir in der ganzen Folge sozusagen alle Realien und Anwendungsarten der menschlichen Erfahrung zu sehen, gruppenweise zugeordnet den Kategorien der fünf Sinne'. Es mag reizvoll sein, im Rahmen eines Symposium über ätherische Öle sich dem '*Geruchsinn*' (V) hinzuwenden. Duftstoffe und die sie liefernden Pflanzen und Tiere sind in überwältigender Fülle dargestellt. Ein Putto reicht einer schönen Nackten Blumen. Sie riecht daran. Vor ihren Füssen rekelt sich eine Zibetkatze umgeben von Parfümbehältern. Eine Räucherpfanne etwa in der Mitte des Bildes und eine Destillationsanlage im linken Hintergrund realisieren den Duft. *Par Fumo*, durch Rauch! Terzetten und Päonien, Rosen und Citrusblüten und viele andere Blüten, Symbolpflanzen des Mittelalters wie Iris, Lilie und Akelei, bedecken eine malerische Wiese, den zusammenfassenden Untergrund dieses Bildes.

Diese Blumenfülle kennen wir von vielen Bildern Jan Breughels d.Ä. Zum einen von seinen Blumenstücken. In der '*Blumengarbe in einem Holzkübel*' (VI) treten uns botanisch exakt in den Details, künstlerisch arrangiert und trotz der Statik des Strausses ein Moment der Bewegung darstellend Blumen über Blumen entgegen. Als wohl auch damals schon obsolete Arzneipflanzen tauchen die alten Sinnbilder des Mittelalters wieder oder immer noch auf: Die Kaiserkrone (*Fritillaria imperialis*) wird von Iris, Madonnenlilie und Rosen auf der einen, von Päonie, Akelei und Feuerlilie auf der anderen Seite eingerahmt.

Dieser Symbolcharakter der Pflanzen, der sich ja bis in unsere Zeit erhalten hat, manifestiert sich auch in einem der wohl schönsten Beispiele meisterlicher Kooperation, in der '*Madonna im Blumenkranz*' (VII) von Jan Breughel d.Ä. und P.P. Rubens. Die Putten und Maria mit dem zur Verehrung entgegengereichen Jesuskind malte Rubens, die Blumengirlande, die sicherlich mehr bedeutet als eine gefällige Dekoration, setzte Breughel hinzu. Erinnert dieses Bild nicht an einen *Hortus conclusus* barocker Art? 'In dem ganzen variierenden Werk bildet der still blühende Märchenkranz gleichwohl eine klösterlich fromme, in sich geschlossene Welt. Die nur dienende leise Blumenmalerei hat sehr viel mehr Religion als die gekonnte Sinnlichkeit der frommen Aktmalerei' (12). Und auch hier ist der Bezug zur Symbolik mittelalterlicher Tafelmalerei unübersehbar: Die Christus- und Mariensymbole Lilie und blaue und gelbe Iris in unmittelbarer Nachbarschaft der Häupter von Mutter und Kind. Die stachelbewehrte Centifolie (*Rosa centifolia*) und die glühendrote '*Rose ohne Dornen*', die Päonie, setzen das Gegengewicht zu Füssen von Mutter und Kind. Symmetrisch links und rechts, unten zwischen den Beinen der Putten hindurchwachsend die Stockrose, Symbol des christlichen Heils, die Bitte um Vergebung und deren Gewährung anzeigend. Und schliesslich als weitere, damals durchaus genutzte Arzneipflanzen vor der (vom Betrachter aus gesehen) linken unteren Putte ein Ast des Schneeballs (*Viburnum opulus*) und hinter ihrer Hüfte Blüten und Blätter des Weissdorns (*Crataegus* species). Ich weiss nicht, inwieweit hier Begriffe wie Heil und Heilung, Ästhetik und dekoratives Pflanzenbeiwerk miteinander verschmolzen sind, bewusst oder unbewusst.

Ein besonders reizvolles Kapitel zum Thema '*Arzneipflanzen in der Symbolik*' ist die Verwendung von Früchten, die ja früher fast alle medizinisch genutzt wurden, als Sinnbilder. Dies gilt ganz besonders für Früchte als Sujets niederländischer Stilleben (13).

Nicht jeder Apfel und jede Nuss auf diesen Gemälden muss eine unmittelbare Bedeutung haben. Oft erschliesst sie sich erst aus dem Bildzusammenhang und dem Ambiente des Werkes. In der Mehrzahl der Beispiele aus dem 17. und 18. Jahrhundert, oft auch des 19. und 20. Jahrhunderts lässt die bewusste Verwendung und Anordnung von Früchten auf die besondere Absicht des Künstlers, den symbolischen Charakter miteinzubeziehen, schliessen. Gerade die Symbolik der Frucht, kann sehr vielschichtig sein und sich in verschiedenen Ebenen abspielen. Vordergründig ist jede Frucht ein Fruchtbarkeitssymbol. Umgekehrt können verschiedene Fruchtarten mit gleichen Eigenschaften Sinnbild für ein Ereignis sein. So stehen viele Früchte, wie z.B. der Granatapfel, die Brombeere, Johannisbeere, Rote Traube, Kirsche, mit rotem Saft für das Blut Christi oder der Märtyrer. Schliesslich kann jede Frucht ein Sinnbild *eigener* Art sein. Sie kann eine symbolische Bedeutung haben, die aufgrund ihrer besonderen Signatur erkennbar ist oder aber erst aufgrund von Hinweisen aus der Literatur erschlossen werden muss.

Interessanterweise hat sich die Bedeutung der symbolischen Charaktere von Früchten und auch Blumen in einigen Fällen im Verlauf der Zeit verändert. So begründete sich die Symbolik des Mittelalters fast ausschliesslich aus dem Christentum. Die Hauptquellen für Deutungen waren zum einen das Alte und Neue Testament, zum anderen die mittelalterlichen naturwissenschaftlichen Schriften z.B. von Albertus Magnus (1193-1280) und Mystik und Dichtung der Hildegard von Bingen (1098-1179). In der Renaissance, im Barock und im Klassizismus ist der Einfluss der Antike, aber auch der des täglichen Lebens, der täglichen Erfahrungen und der Umwelt, z.B. in Sprichwörtern und Redensarten, auf die Verwendung von Symbolen nicht zu übersehen. Natürlich vermischen sich oft beide Bereiche und machen so die Deutung schwierig und oft widersprüchlich. Denn auch dies ist eine Eigenschaft der Pflanzen- und damit der Fruchtsymbolik: häufig ist sie zweideutig; gut und böse, alt und jung, schön und hässlich können von dem selben Objekt ausgedrückt werden.

Ein sehr schönes Beispiel für das eben Dargelegte ist die Symbolik des Pfirsichs, eine auf niederländischen Stilleben immer wieder gemalte Genuss-, aber auch Medizinalpflanze. Ihrer hat sich in vielfältiger Weise Willem van Aelst (ca. 1625-1683) angenommen. Zusätzlich zu Pfirsichen sind es vor allem Trauben, die ihn in ihrer gegensätzlichen Grösse und Beschaffenheit der Oberflächen fasziniert haben. In seinem 1674 gemalten '*Pfirsichstilleben mit der Maus*' (VIII) wird diese Meisterschaft deutlich. Das dunkle Lila der Samttischdecke wird von der Oberflächenstruktur her auf die seidige Haut der beiden Pfirsiche übertragen. Das Violett derDecke findet sich wieder in den teilweise bereiften, teilweise blanken Weintrauben. Aber auch die rechts unten sitzende Maus hat in ihrem schimmernden bräunlichen Fell, in den nackten Ohren und in dem haarlosen Schwanz Spuren dieser unnachahmlichen Lila-Tönung. Der Symbolcharakter dieses Bildes ist gross. Im Mittelpunkt liegen die Pfirsiche; zunächst '*Äpfel*' im erweiterten Sinn und damit Symbol für das

Unterscheidungsvermögen von Gut und Böse, vor allem aber auch für das Gute und Böse, besonders im Sinne von Wollust, selbst. Verschmelzen hier antike Attribute der Venus mit christlichen der Maria? Weist hierauf die Verwendung des Pfirsichs als Sinnbild der Dreifaltigkeit, der Tugend und Erlösung aufgrund der Dreiteilung in Fruchtfleisch, Stein und Kern?

Den Dualismus und die Ambivalenz in der Deutung des Pfirsichs finden wir auch in der Weintraube wieder. So ist sie Sinnbild Mariens und der Heiligen Kirche, vor allem aber Symbol Christi: Im Johannesevangelium (14,7; 15,1; 15,5) bezeichnete er sich selbst als wahren Weinstock. Beim Letzten Abendmahl nennt er den Wein sein eigen Blut (Matth. 26,28). Schon im Alten Testament wurde die Weinrebe dem Frieden und dem Wohlstand des Volkes Israel, ja Israel selbst gleichgesetzt (Hosea 10,1). Symbol der Fruchbarkeit, aber damit einhergehend auch der Erotik war die Traube aber immer schon ausserhalb des christlich-jüdischen Kulturkreises gewesen. Ist die Darstellung eines bacchantischen Festgelages ohne Weintrauben denkbar?

Schliesslich soll in dem Stilleben von Willem van Aelst der ambivalenten Symbolik der beiden Fruchsorten wohl ein in seiner Bedeutung eindeutiger diagonaler Schwerpunkt gesetzt werden: Die Maus als Sinnbild irdischer Gier und der Tagfalter als Symbol der erlöst zum Himmel aufsteigenden Seele (13).

Lassen Sie mich an einem letzten Beispiel die Symbolik von Arzneipflanzen aufzeigen: am Granatapfel (*Punica granatum*). Ich möchte an dieser Frucht deutlich machen, wie über Jahrtausende hinweg ein Symbolcharakter erhalten bleiben kann. Schon in der Antike galt er seiner vielen Samen wegen als Symbol der Fruchtbarkeit und der Unsterblichkeit. So stand er in der ägyptischen aber auch in der chinesischen Kunst in vielfältiger Weise Modell. Beim Volke Israel spielten die Granatäpfel eine bedeutsame Rolle. Sie zierten die Gewänder der Priester und die Vorhänge des Tempels, sie waren eingemeisselt in die Schmuckbänder der steinernen Gotteshäuser. Der Granatapfel symbolisierte Gesetz und Thora, und damit Gottes Wort und Wille, das Höchste, was der Alte Bund kannte.

Die christliche Kunst übernahm die Symbolik des Granatapfels. Wohl vom Hohen Lied (4,3) abgeleitet – hier wird die Menge der Tugenden mit den vielen Kernen des Granatapfels verglichen – entstanden auf Christus, Maria und die Kirche bezogene Symbolbedeutungen. Das Tafelbild von Filippino Lippi (ca. 1457-1504) '*Maria mit dem Granatapfel*' (IX) und das 1519 von M. Grünewald (eigentlich: Mathis Gothart Nithart, 1408?-1528) geschaffene Altarbild der '*Stuppacher Madonna*' (X) belegen das. Auf beiden Bildern hält die Himmelskönigin ihrem Sohn einen Granatapfel entgegen. Diese Mariensymbolik steht unter anderem für überströmende Liebe, für Caritas, für unendliche Tugenden. Als Christussymbol verkörpert der Granatapfel wegen seines roten Fruchtinner nicht nur das Blut Christi, sondern auch als Frucht, die eine Krone trägt, sowohl die Kirche als auch die Welt. Ein aufgebrochener Granatapfel, wie er in verschiedenen Niederländischen Stilleben zu finden ist, z.B. bei Pieter de Ring (1615-1660) '*Kartusche mit Blumen und Früchten* (XI), bei Abraham Breughel (1631-1697) '*Frau, die die nach den Früchten greift*' von 1669 (XII), bei dem niederländischen Rokokomaler Jan van Huysum (1682-1749) '*Stilleben mit Blumen und Früchten*' (XIII), zeugt von der Liebe Christi, die allen Menschen offen steht (nach 13). Hier wird aber bereits die Schwierigkeit der symbolischen Deutung sichtbar. Der Granatapfel vereint die Gegensätze süss und herb. So sind in ihm Gut und Böse, Strenge und Milde, Caritas und Eros eingeschlossen. Ich meine, diese Diskrepanz wird schon im Hohelied Salomos (4,13,16) erkennbar. Mit anderen Pflanzen wird der Granatapfel zusammen mit dem Duft in den Mittelpunkt der Liebe gestellt: Gott bzw. Christus als Bräutigam und Israel bzw. die Ecclesia als Braut. Damit wird auch die Rolle dieser Frucht im Spannungsfeld der Geschlechter deutlich.

Gerade der Granatapfel, der in so vielseitiger Weise zu deuten ist, und der in so vielfältiger Weise in der Kunst, auch in der niederländischen Kunst als Objekt genutzt wird, gibt mir Gelegenheit zu zeigen, dass auch heutige Maler pflanzliche Symbolik einzusetzen verstehen. Eberhard Schlotter hat in seinem 1977 entstandenen Halbakt '*Nocturno*' (XIV) das Weiblich-Erotische, Schwebende und für uns Männer oft nicht Fassbare dargestellt. Eine reizvolle, begehrenswerte Frau beherrscht das Bild. Sie wird durch zwei Arzneipflanzen symbolisiert: die weiss blühende *Datura*, deren Duft – ich zitiere den Surrealisten André Breton – '*das wilde Entzücken der vollkommenen Gegenwärtigkeit darstellt*', und die geöffnete Frucht der vieltausendjährigen Symbolpflanze *Punica granatum*, die hier das erotische Fluidum der Frau versinnbildlicht.

Der Hinweis auf die vieltausendjährige Symbolik des Granatapfels im Zusammenhang mit dem Spannungsfeld von Caritas und Eros, in der Ambivalenz der Fähigkeit, Gut und Böse zu erkennen, soll meine Exkursion durch diesen niederländischen Arzneipflanzengarten sui generis beenden. Wir haben diese und jene Pflanze betrachtet. Ich habe versucht, die Symbolik einzelner Pflanzen aus ihrer Zeit heraus, aber auch in ihrer Entwicklung in den Jahrhunderten aufzuzeigen. Ich habe versucht, Sie wieder oder wieder *intensiver* Sehen zu lehren.

Hat Josef Pieper recht, wenn er meint, viele von uns hätten das Sehen verlernt durch zuviel Sehen? Können wir nur noch plakative aufnehmen: einfarbige Blütenfelder, eintönige Zierrasen, einsame Flächen moderner Künstler? Verliert der Mensch der Gegenwart mit dem Unvermögen, Details zu erkennen, auch die Empfindsamkeit und die Empfindlichkeit für kleinste Veränderungen? Sehen sollte kreativ sein. Es sollte sensibilisieren und Denkanstösse geben. Sehen kann beglücken. Sehen kann verzaubern. Müssen wir das Sehen wieder erlernen? Dieses Sehen im Detail ist Eigenart und Eigenschaft des Augentieres Mensch. Gerade als Naturwissenschaftler sollten wir uns mit Hilfe der Kunst das Glück gönnen, mit Verstand und mit Gefühl zu sehen.

LITERATUR

(1) Behling L. 1967, Die pflanzen in der mittelalterlichen Tafelmalerei, Köln/Graz, Böhlau-Verlag.

(2) Czygan F-C. 1981, Ätherische Öle und Duft – kulturhistorisch betrachtet, Pharmazie in unserer Zeit 10:109-121;
Czygan F-C. 1984, 4000 Jahre Ätherische Öle – eine kunst- und kulturhistorische Exkursion, Ärztez. f. Naturheilverf. 25:500.

(3) Behling L. 1964, Die Pflanzenwelt der mittelalterlichen Kathedralen, Köln/Graz, Böhlau-Verlag.

(4) Czygan F-C. 1983, Ätherische Öle, Duft und Literatur: ein naturwissenschaftlich-poetischer Dreiklang, Kneipp-Physiotherapie 3:2-5.

(5) Lehner E, Lehner J. 1960, Folklore and symbolism of flowers, plants and trees, New York,

(6) Schmidt H, Schmidt M. 1982, Die vergessene Bildersprache christlicher Kunst, München, CH Beck.

(7) Sachs H, Badstübner E, Neumann H. o.J., Erklärendes Wörterbuch zur christlichen Kunst, Hanau, W Dausien-Verlag.

(8) Leyser H (Herausg.). 1838, Deutsche Predigten des 13. und 14. Jahrhunderts, Quedlinburg/Leipzig.

(9) Würzburg K von. 1966, Die goldene Schmiede ('Altdeutsche Wälder' hrsg. von den Brüdern Grimm, Frankfurt, 1815), Reprint Darmstadt.

(10) vgl. (7), Seite 285.

(11) Kauffmann H. 1943, Die Fünfsinne in der niederländischen Malerei des 17. Jahrhunderts, in: Kunstgeschichtliche Studien. Festschrift für Dagobert Frey (Tintelnot H, Hrsg.), Breslau.

(12) Winkelmann-Rhein G. 1979, Blumen-Breughel, Köln, DuMont Buchverlag.
zu J. Breughel d.Ä. vgl. auch
Baumgart F. 1978, Blumen Breughel, Köln, DuMont Buchverlag.

(13) Segal S. 1983, Niederländische Stilleben, Katalog zur Ausstellung, Braunschweig, Herzog Anton Ulrich-Museum.

(7a) Fritz R. 1952, Aquilegia, die symbolische Bedeutung der Akelei, Wallraf-Richartz-Jahrbuch, Köln;
Löber K. 1977, Pflanzensymbolik der mittelalterlichen Tafelmalerei mit besonderer Berücksichtigung der Akelei, Symbolon N.F. Heft 3.

(I) *Noli-me-tangere-Altar*: ca. 1425, Meister des Göttinger Barfüsser-Altars; Staatsgalerie, Stuttgart.

(II) *Maria im Rosenhag*: 1473, Martin Schongauer (1450-1491); Münster St. Martin, Colmar, Frankreich.

(III) *Maria im Rosenhag*: um 1448, Stephan Lochner (1400/15-1451/52); Wallraf-Richartz-Museum, Köln.

(IV) *Portinari-Altar*: um 1473-1475, Hugo van der Goes (1440/45-1482); Uffizien, Florenz.

(V) *Allegorie der Fünf Sinne* (hier: *Der Geruch*): 1617-1618, Jan Breughel d.Ä. (1568-1625); Prado, Madrid.

(VI) *Blumengarbe in einem Holzkübel*: um 1610?, Jan Breughel d.Ä.; Alte Pinakothek, München.

(VII) *Madonna im Blumenkranz*: 1618-1620, Peter Paul Rubens (1577-1640) und Jan Breughel d.Ä.; Alte Pinakothek, München.

(VIII) *Pfirsich-Stilleben mit der Maus*: 1674, Willem van Aelst (ca. 1625-ca. 1683); Privatsammlung.

(IX) *Maria mit dem Granatapfel*: Filippino Lippi (1457?-1504); Louvre, Paris.

(X) *Stuppacher Madonna*: 1519, Matthias Grünewald – eigentlich: Mathis Gothart Nithart – (ca. 1480-1528); Stuppacher Kirche bei Bad Mergentheim, BRD.

(XI) *Kartusche mit Blumen und Früchten*: Pieter de Ring (1615-1660); Gallerie Müllenmeister, Solingen, BRD.

(XII) *Frau, die nach Früchten greift* (Ausschnitt: *Granatäpfel*): 1669, Abraham Breughel (1631-1697); Louvre, Paris.

(XIII) *Stilleben mit Blumen und Früchten*: Jan van Huysum (1682-1749); Rijksmuseum, Amsterdam.

(XIV) *Nocturno*: 1977, Eberhard Schlotter (1921-); Privatbesitz.

GPSR Compliance

The European Union's (EU) General Product Safety Regulation (GPSR) is a set of rules that requires consumer products to be safe and our obligations to ensure this.

If you have any concerns about our products, you can contact us on ProductSafety@springernature.com

In case Publisher is established outside the EU, the EU authorized representative is:

Springer Nature Customer Service Center GmbH
Europaplatz 3
69115 Heidelberg, Germany

Batch number: 10607222

Printed by Printforce, the Netherlands